RUBBER COMPOUNDING

RUBBER COMPOUNDING

PRINCIPLES, MATERIALS, AND TECHNIQUES

SECOND EDITION

FRED W. BARLOW
Rubber Consultant
Stow, Ohio

CRC Press
Taylor & Francis Group
Boca Raton London New York

CRC Press is an imprint of the
Taylor & Francis Group, an **informa** business

CRC Press
Taylor & Francis Group
6000 Broken Sound Parkway NW, Suite 300
Boca Raton, FL 33487-2742

First issued in paperback 2019

© 1993 by Taylor & Francis Group, LLC
CRC Press is an imprint of Taylor & Francis Group, an Informa business

No claim to original U.S. Government works

ISBN-13: 978-0-8247-8968-8 (hbk)
ISBN-13: 978-0-367-40231-0 (pbk)

Library of Congress Cataloging-in-Publication Data

Barlow, Fred W.
    Rubber compounding : principles, materials, and techniques / Fred
W. Barlow. -- 2nd ed.
        p. cm.
    Includes bibliographical references and index.
    ISBN 0-8247-8968-7
    1. Rubber. 2. Elastomers. 3. Rubber chemicals.   I. Title.
TS1890.B268   1993
678'.23--dc20   93-24756
                                                                          CIP

Visit the Taylor & Francis Web site at
http://www.taylorandfrancis.com

and the CRC Press Web site at
http://www.crcpress.com

Preface to the Second Edition

Since the first edition was published, there have been many developments that today's compounders should know about if they are to ply their profession successfully. I have tried to incorporate these in this second edition.

For example, the present edition gives an introduction to thermoplastic rubbers. These hybrids of plastic and rubber have a growing place in factories that heretofore dealt strictly with rubber goods. Again competition in marketing rubber products has never been fiercer so methods of waste minimizing are described. I also outline how to conduct in-plant research more economically by using simple statistically designed experiments.

There has been a steady increase in environmental regulations that affect compounders. I have tried in this edition to identify materials considered toxic and to show how waste can be disposed of in compliance with these regulations. Finally, brief descriptions of the two federal acts that relate most to compounding—The Resource Conservation and Recovery Act (RCRA), 1976, and the Superfund Amendments and Reauthorization Act (SARA), 1986—are described along with the reporting that is necessary.

My thanks to those who called my attention to inaccuracies in the first edition.

Fred W. Barlow

Preface to the First Edition

Long ago someone described a book's preface as the author's excuse for writing the book. So I'll start by giving mine. Currently, there are few books that deal exclusively with rubber compounding. There are many courses on rubber compounding available on those in the rubber industry. These courses consist mostly of lectures on rubber-compounding materials by experts, usually from supplier companies. Although the material presented is generally excellent and up to date, deficiencies or difficulties encountered with the ingredients may be understated, and there is not an ongoing attempt to weave the information together to show how compounds are built up. This book is an attempt to fill those voids.

The author has at one time or another worked with the compounding of all the polymers described here. But obviously no comprehensive book on compounding can be written from the experience and remembrances of one person. Accordingly, the voluminous technical literature has been freely consulted and used. A constant problem has been judging what would be most useful to include. Certainly carbon blacks and accelerators cannot be excluded but what about fire retardants and mold release agents? This book contains brief notes on fire retardants but mold release agents have been omitted.

It is inevitable in a book of this kind that errors will creep in and for these the author takes sole responsibility. Although rubber technology is not changing as fast as some other technologies, it *is* changing and this means the information will change in time. It cannot be too strongly urged that compounders keep in close touch with their suppliers, environmental officials, customers, and the technical literature to make sure they stay close to the "cutting edge."

Successful compounding!

Fred W. Barlow

Contents

1

Introduction to Compounding

I. SCOPE

This book was written to provide a guide to modern compounding, describing the materials involved as well as why and how they are used to build up a compound. Many technologists would agree that rubber compounding is still more an art than a science, so the emphasis is on simplicity and common usage and understanding. However, where scientific study is providing answers, pertinent material is included or noted. The book is fashioned to serve the needs of the beginning compounder as well as to provide a useful reference for more experienced readers or those who serve the rubber goods industry—for example, suppliers.

The compounding, mixing, and testing of rubber stocks covers a formidable array of materials and techniques. To attempt to cover them all would result in a cumbersome and disjointed work. Accordingly this book is concerned only with five general-purpose rubbers and three specialty rubbers. The general-purpose rubbers are natural, styrene-butadiene, polybutadiene, butyl, and ethylene propylene. They are called general-purpose rubbers because of their usage in a wide variety of products. The three specialty rubbers are neoprene, butadiene-acrylonitrile, and polysulfide rubbers. These are primarily chosen for their resistance to solvents, especially petroleum-derived solvents, and find use in such applications as gasoline and paint spray hose tubes. These eight rubbers account for more than 96% of the world's consumption.

In addition, two other rubber-based materials are reviewed. A class of material called thermoplastic rubbers is of increasing importance. For our purpose these might be called blends or alloys of a rubber with a plastic. Unlike other rubbers, they soften considerably on heating. They are included here because rubber is an integral part of their structure, and because they are compounded with other materials and often can be run on the same equipment. Chapter 6 contains a section on rubber reclaims and reground scrapped rubber goods. Such products have increased in value in the recent past because they can be recycled.

Although they have different shades of meaning, the terms *polymer*, *elastomer*, and *rubber* are used interchangeably in this book simply to avoid repetition. No matter what term is used, the compounder has a mountain of these to work with. The International Rubber Study Group estimated that 14.8 million metric tons of rubber was produced in 1990. Of this amount, 4.93 million tons or 33.25% was natural, the remaining 9.89 million tons, 66.75%, synthetic. Production of all rubbers usually follows consumption closely, and growth in the last 5 years has been slow. Production of all rubbers in 1990 was only 11% above that of 1985.

The variety of rubber compounds used is enormous. One company with which the author was associated—a mechanical rubber goods manufacturer—had formulas for 800 compounds, although normal inventories consisted of only a fraction of that number. Considering the variety of rubber products and the number of manufacturers, it would appear that thousands of compounds are available, although not in regular use. Whether such large numbers are necessary is questionable. The smaller the number of compounds a manufacturer can offer to satisfy his customers, the greater the savings in cost accounting, and the greater the economies that can be achieved in terms of longer mixing runs, and lower costs associated with warehousing, loss from storage hardening, and quality control.

II. DEMANDS AND REWARDS

Good compounding means the development of formulas that are environmentally safe, factory-processible, cost-competitive with other compounds in the same applications, and able to provide a satisfactory service life. These requirements put considerable demands on the compounder, who must have extensive knowledge of materials at least as diverse as carbon black and sulfur, zinc oxide, and ester plasticizers. There must be sufficient familiarity with rubber processing equipment for reasonable assurance that compounds can be processed without difficulty. Costs, always a major concern, frequently are expressed as cents per cubic inch or a suitable metric volume measure. This is because in many instances there is a standard practice of using a certain thickness or volume of rubber compound, like the cover on a belt.

Since the first edition of this book, two areas of interest to compounders have grown in importance, namely their responsibilities in environmental areas and their own professional liability. These rising concerns cannot be ignored in the United States, and they have larger or smaller counterparts in many other countries. Environmental regulations affecting the rubber goods industry are reviewed in Chapter 20 (Waste Reduction and Disposal). The complexity of the regulations is attested by the growth in the number of environmental consultants. In many large plants environmental compliance is the responsibility of an official, who is not the rubber chemist. The compounder's knowledge may be invaluable to the designated officer, however, and in smaller plants the compounder may be the person most qualified for the job. Compliance is very important when a severe penalty can be applied against a plant manager who allows such a small fault as a janitor throwing an oily rag into the office's paper waste. (Oily rags constitute a hazardous substance.)

Compounders are professionals and as such have professional liability in the same way as an engineer or architect is professionally liable for the faulty design of a machine or building. If one is not self-employed, it would be prudent to learn whether the terms of employment shield one from claims of damages for alleged negligence. For the compounder who is self-employed, say as a consultant in rubber product design, there are problems. Professional liability insurance is expensive and difficult to obtain, yet the cost of defending one's self in court can be considerable. Protection without insurance is possible, however, and some approaches are described in a book by H. Streeter [1].

Although the demands on the compounder are heavy, the rewards can be satisfying. There are no safety hazards, and the customer is satisfied with a good quality product, and with low scrap losses, reduced energy consumption from less reworking of bad compound, and a cost-competitive compound, factory profit possibilities are enhanced—all of which give a deep sense of accomplishment to the compounder.

III. CLASSIFICATION OF MATERIALS

It is customary when discussing compounding to classify the materials used by the function they serve. Some materials, such as zinc oxide, can behave in more than one way. This book uses the following classification:

1. Elastomers
2. Vulcanizing or crosslinking agents
3. Accelerators
4. Activator/retarders
5. Process aids
6. Softeners and plasticizers
7. Reinforcers/fillers—black

Table 1.1   Conveyor Belt Cover Compound

Material	Parts	Functional class
Natural rubber (SMR CV)	100.0	Elastomer
N 220 carbon black	45.0	Black reinforcer
Zinc oxide	4.0	Activator
Stearic acid	2.0	Activator
Rubber process oil	4.0	Softener
Agerite Resin D[a]	2.5	Age resister
Paraffin wax	1.0	Age resister
CBS[b]	0.5	Accelerator
Sulfur	2.5	Vulcanizing agent
	161.5	

[a]R.T. Vanderbilt Company trade name.
[b]N-Cyclohexylbenzothiazole-2-sulfenamide.

   8.   Reinforcers/fillers—nonblack
   9.   Age resisters
  10.   Miscellaneous materials

A chapter is devoted to each class in turn (several in the case of elastomers), and the remainder of the book is directed to how the information is used for commercial compounding.

A typical compound for a first grade conveyor belt cover is shown in Table 1.1. Compounds do not always require material from each of the foregoing classes. The natural rubber compound used as an example indicates in a very general way the amount of each kind of material usually used. Certain sorts of ingredients, such as age resisters, are available in many nongeneric products, and such trademarked materials are often specified. A common convention among compounders is to base formulations on 100 parts of elastomer and to list the elastomer first. Frequently more than one rubber is used. Less widely used but helpful is the practice of placing the names of the accelerators and vulcanizing agent(s) at the end of the compound, to expedite the identification of the curative system.

English, metric, and SI units are commonly encountered in the literature of rubber technology. Appendix 3 (Weights, Measures, and Conversion Factors) gives the various relations and illustrates a method for more "fool-proof" conversions from units of one type to another.

IV.   TRAINING FOR COMPOUNDING

Training for rubber compounding is usually on-the-job training in the United States, with entry level jobs often taken by science or engineering graduates. Not infrequently the latter are from chemistry or chemical engineering courses rather than physics. This is somewhat strange, for although chemical problems exist in such areas as vulcanization and aging, it is the physical properties of rubber that are most appreciated in its service applications.

In some countries rubber technology is taught as a college level course. One of the best known facilities for such study is the London School of Polymer Technology (formerly known as the National College of Rubber Technology) in England. Because of its reputation, the school attracts students worldwide; it offers B.Sc. and M.Sc. degrees, as well as a lower level associateship. As an illustration of what compounding is about and the knowledge needed to be a rubber technologist, consider two examinations for the associateship. The first, Paper III: R Rubber Technology 2, is largely concerned with compounding and laboratory testing; the second, Paper II R: Rubber Technology I, deals with rubber product design and factory operations. Both tests are reproduced with the permission of the London School of Polymer Technology, London, England.

Graduateship of the Plastics and Rubber Institute and Associateship of the London School of Polymer Technology Examination, 1986

Paper IIIR   Rubber Technology 2

Monday 2nd June 1986   2.00–5.00 PM

*Answer FIVE questions
*ALL questions carry equal marks

1. Give a detailed account of the factors you would consider when choosing between natural rubber (NR) and styrene-butadiene rubber (SBR) as the base polymer for a rubber product.

    Select one product for which you think NR is to be preferred, stating your reasons, and design a compound suitable for the application.

2. Discuss the relationship between the properties and structure of butyl rubber (IIR).

    What are the main technological problems associated with the use of butyl rubber, and how may they be overcome?

3. Discuss the use of silica fillers in the compounding of rubbers.
    Describe the advances which have been made to extend their use.

4. Give an account of the technological factors which have prevented the widespread use of organic peroxides as vulcanising agents.

    Discuss the developments which have occurred which may lead to greater use of peroxide vulcanising agents in the future.

5. Discuss the problems which exist in the use of antioxidants for the protection of rubber vulcanisates.

    Describe the progress which has been made towards overcoming these problems.

6. Describe how conditions at the test machine/test piece interface influence the design of standard compression stress/strain and compression set tests. A cylindrical elastomer test piece of 20 mm height and 45 mm diameter is subjected to a compressive force of 400 N normal to its ends and under non-slip conditions. Calculate the compression strain in the test piece if the compound has a Young's modulus of 4.5 MPa and the value of the compression factor, k, is 0.56.

7. How is the tearing energy of an elastomer related to its surface free energy? Describe a method by which the tearing energy of an elastomer compound might be determined. Your account should include sketches of typical force/ deformation curves and their interpretation, and the derivation of any formulae quoted.

8. Routine process control operations in rubber factory mixing departments are now carried out, almost entirely, by means of curemeters. Describe, briefly, a curemeter suitable for this application, showing how relevant control data might be obtained from typical cure traces of the plateau, reverting and "marching modulus" types.

Paper II R Rubber Technology 1

Friday 30th May 1986   2.00–5.00 PM

*Answer FIVE questions
*ALL questions carry equal marks

1. Discuss the role of surfactants in the production, storage and use of polymer latices.

2. Describe three methods of producing dipped films from polymer latices and discuss the advantages and limitations of each method.

    Outline the fundamental criteria which control the deposition of a uniform latex film onto a dipping former.

3. Explain the meaning of the term "neutral angle" as applied to the reinforcement of hose and derive its value.

    Derive an equation for the bursting pressure of wrapped delivery hose and show how this can be used to determine the number of plies required in a hose construction.

4. Describe and explain the differences in design between a radial tyre and a cross-ply tyre.
   Explain how these differences lead to the superior performance of the former.

5. Compare and contrast the properties of cotton with those of viscose rayon. How may the properties of viscose rayon be changed by modifications to the method of production?

6. What are the advantages and limitations of ram extruders compared with screw extruders?
   Describe two uses of the ram extruder in the rubber industry, indicating the reasons for the use of this type of machine.

7. "The process of calendering is basically simple, but has been refined to the point where it is now possible to produce sheets of very uniform gauge provided that one is prepared to pay for it."
   Discuss this statement, describing the ways in which calenders and calendering techniques have developed to produce more uniform products.

8. Describe how the transmissibility of rubber is dependent on the ratio of imposed frequency to natural frequency.
   What conclusions can be drawn on the use of rubber for the isolation of vibrations?

REFERENCE

1. Streeter, H., *Professional Liability of Architects and Engineers*, Wiley, New York, 1988.

Elastomers: Natural Rubber

I. SCOPE

Rubber is to rubber compounds as flour is to bread: although the choice and amount of additives can give considerable variation to the end product, the main characteristic determinant is the kind of flour (or rubber) used. This chapter examines briefly the chemical structure properties of polymers used in compounding. Detailed consideration is then given to the original rubber, natural rubber. The American Society for Testing and Materials (ASTM) has adopted convenient abbreviations for commercial elastomers: natural rubber is simply NR.

The term *rubber* is difficult to define. A somewhat cumbersome definition is given in ASTM D 1566-88b, Standard Terminology Relating to Rubber [1]:

> rubber—a material that is capable of recovering from large deformations quickly and forcibly, and can be, or already is, modified to a state in which it is essentially insoluble (but can swell) in boiling solvent, such as benzene, methyl ethyl ketone, and ethanoltoluene azeotrope. A rubber in this modified state, free of diluents, retracts within one minute to less than 1.5 times its original length after being stretched at room temperature (18 to 20°C) to twice its length and held for one minute before release.

Technical definitions of rubber are of interest not only to compounders. Customs officers, for example, require a precise definition of rubber to ensure that the appropriate import duty, if any, is collected. We do not need such high definition here.

II. CHEMICAL STRUCTURES

As the name implies, reclaimed rubber is derived from such materials as tire treads. Although it can be a very useful rubber in low cost compounding, it is not a virgin polymer and its structure is not discussed here. Of the eight principal rubbers considered in this text, five are hydrocarbons consisting solely of carbon and hydrogen. They are natural, styrene-butadiene, butyl, polybutadiene, and ethylene propylene rubbers. These five are classed as general-purpose rubbers because of their large use in high volume rubber goods such as tires, hoses, and belting. The remaining three are largely carbon and hydrogen, although butadiene-acrylonitrile rubber contains nitrogen as well, neoprene has chlorine as its third element, and the polysulfide rubbers have sulfur. These special-purpose rubbers are used where such properties as high resistance to oil or solvents is required.

Polymers are high molecular weight compounds made from low molecular weight building units called monomers. There may be 1000–20,000 repeating units of the monomer. If the polymer consists solely of one monomer, it is called a homopolymer; if it has two species, it is a copolymer; if three, a terpolymer. Examples are, respectively, natural rubber (polyisoprene), butadiene-styrene rubber, and ethylene propylene terpolymer, where the third monomer may be dicyclopentadiene.

In the early days of synthetic polymer manufacture, bulk polymerization was the method used and monomer was polymerized in bulk, often with the aid of a catalyst. This process was superseded by emulsion polymerization, the main process used today, although solution polymerization is of growing importance. In the emulsion process the monomers are dispersed in water with a surface active substance such as soap and a polymerization initiator. Other materials are usually included, but the three above are the most important. The result of the reaction is a dispersion of minute particles of polymer in an aqueous medium called latex. The minute particles are coagulated and dried to form the rubber product. Compared to bulk polymerization methods, the emulsion method results in a better product at a quicker rate. In solution polymerization the monomer is polymerized in a suitable solvent. Butyl rubber is produced in this way.

Historically, polymerization was considered to occur by two methods, addition and condensation. In the first method the monomer units join without any molecule being eliminated; in the second, a small molecule is spun off during each step in the polymer building reaction. Morton [2] has pointed out the flaws in such a division—some so-called polycondensates like nylon 6 could result from addition polymerization. This author recommends that polymerization reactions be classified according to the way in which they affect the molecular size and the size distribution of the final product. If this convention is accepted, there are only two basic processes whereby macromolecules are synthesized: functional group polymerization and chain addition polymerization.

Functional group polymerization has two classes:

$$A—R—A + B—R'—B \longrightarrow A—R—R'—B + AB$$

Polycondensation

$$A—R—A + B—R'—B \longrightarrow A—R—AB—R'—B$$

Polyaddition

where A and B are functional groups that react with each other. Of the rubbers we review, only the polysulfide rubber is made by polycondensation.

Chain addition polymerization occurs with the other rubbers discussed and involves the successive additions of monomers to a growing chain, which is initiated by some active species.

Morton goes on to illustrate how these polymerizations proceed by three possible mechanisms and may involve multiple bonds or rings. For reacting with double bonds the equations are:

Free Radical

Cationic

Anionic

In the first equation the free radical is a carbon atom, which is highly reactive because it has only seven outer electrons (one unpaired) instead of eight. In the second, the propagating center, the carbonium atom, has only six outer electrons and therefore has a net positive charge. Finally, carbanions or carbon anions have a net negative charge because only three of the four valencies are satisfied despite the normal eight outer electrons.

The significant distinction between functional group polymerization and chain addition polymerization is that in the latter each macromolecule is formed by a chain reaction, which is initiated by some activation step.

In polymerization, the double or triple bond between two carbon atoms is broken up by using a catalyst under appropriate conditions. With the bond broken, the monomeric unit is chemically active and can unite with other units to form the polymer. A simple illustration appears in Figure 2.1, which shows ethylene gas transformed into the solid transparent polymer polyethylene. Polymers such as this made from one monomer are called homopolymers. This simple linear polymer already has some complexity: the molecular chains are not all of the same length.

A second complication of structure occurs, still dealing with homopolymers, if there is more than one double bond in the monomer. This is the case when butadiene (C_4H_8) is polymerized to polybutadiene. The monomers may add on not only head to tail but also intermediately, as illustrated in Figure 2.2. To help specify the location of additions, the carbon chain is numbered so that the numbers used are in order and the smallest possible, and the type of addition is identified by the number of carbons involved in the linkage.

A third complexing element in such polymerizations is the structure of the molecule within the addition that is due to geometrical isomerism. When two carbon atoms are joined by a double bond, they are locked together in a rigid structure. If the two groups attached to each of the carbon atoms in such a pair are unlike, they can form a material in which two like groups are on the same side, this is called the *cis* form; if two like groups are on opposite sides, the polymer is said to have the *trans* form. Figure 2.3 shows this phenomenon in

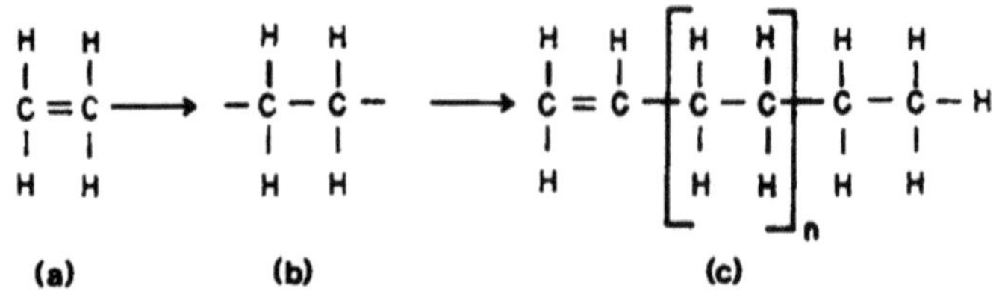

Figure 2.1 Polyethylene preparation: (a) ethylene gas (monomer), (b) active unit, and (c) polyethylene, solid (polymer).

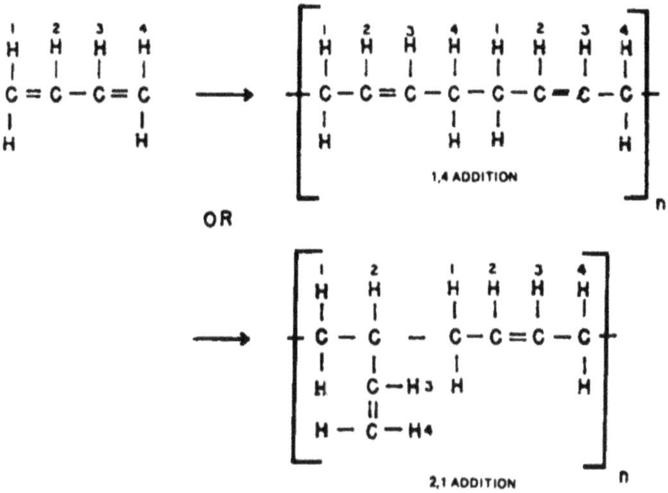

Figure 2.2 Two types of addition in butadiene polymerization.

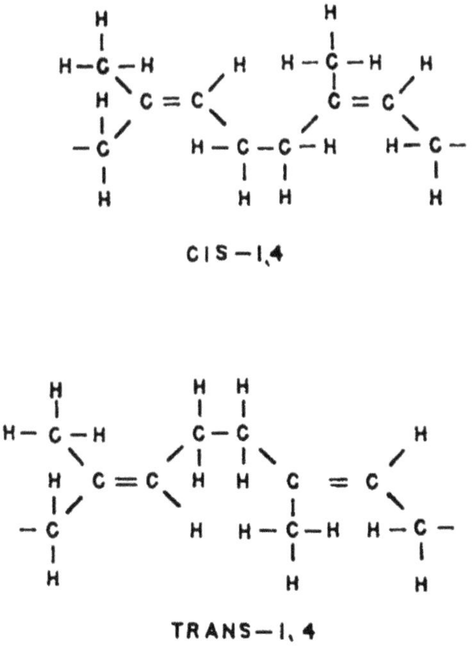

CIS—1,4

TRANS—1,4

Figure 2.3 Stereoisomerism in natural rubber.

natural rubber. Natural rubber consists largely of *cis*-1, 4-polyisoprene. Balata (a tough horny substance used at one time for golf ball covers) is largely *trans*-1, 4.

In the foregoing we have considered only homopolymers, but polymers can be built up from two or three monomers. Styrene-butadiene rubber uses two monomers, ethylene propylene terpolymer three. These polymers can have additional varying properties depending on the ratio of the monomers in the polymer and how the monomer units are placed. Block rubbers, for example, have a chain of one type of monomer units followed by a chain of another, or the monomer units can be randomly mixed as in styrene-butadiene rubbers. In a third kind of arrangement, called graft polymerism, groups of the monomer are suspended from the main chain.

For compounding skill, some understanding of the foregoing polymer properties is required, especially as they relate to the rubbers being used in the plant. Rubber suppliers use advanced compounding techniques to relate as accurately as possible polymer composition and structure to such practical considerations as wet skid resistance in passenger tires.

Certain polymer characteristics are known to be necessary in rubberlike materials. For example, the material must be a high molecular weight polymer with long flexible chains; it must be consistently above its glass transition temperature at room temperature. The glass transition temperature, symbolized by T_g, is the midpoint of the temperature range during which the polymer changes from a rubberlike state to a glassy state, where it is easily shattered. Crystallinity must be low at working temperatures. Finally the material must be crosslinkable; that is, the molecular chains must be able to accommodate the formation of intermediate bonds during vulcanizing. These bonds ensure that a stable, insoluble, three-dimensional network is formed.

III. NATURAL RUBBER

Any review of the major rubbers fittingly starts with natural rubber. It was the first rubber and was unique until the development of polysulfide rubber in about 1927. NR supplies about one-third of the world demand for elastomers and is the standard by which others are judged. It was Charles Goodyear and Thomas Hancock's experiments with natural rubber, sulfur, and heat that led to the discovery of vulcanization, representing the birth, as it were, of compounding.

A. Properties

Chemically, natural rubber is *cis*-1, 4-polyisoprene. A linear, long-chain polymer with repeating isoprenic units (C_5H_8), it has a density of 0.93 at 20°C. Intensive plant breeding has produced a wide range of clonal types whose Mooney viscosity (when freshly coagulated) can vary from slightly below 50 to over 90. Mooney viscosity is a common test for toughness of a rubber; the higher

Table 2.1 Typical Analysis of Natural Rubber

Component	%
Moisture	0.6
Acetone extract	2.9
Protein (calculated from nitrogen)	2.8
Ash	0.4
Rubber hydrocarbon	93.3
	100.0

the value, the more resistant to deformation. Paralleling this Mooney range is a very wide range of molecular weights. Since, however, rubber in commerce is a blend of rubbers from various clones, the spread is less. It has been estimated that a random blend would have a weight average molecular weight of perhaps 2 million and a number average molecular weight of 0.5 million.

Due to the blending mentioned above, the Mooney viscosity for commercial rubbers is approximately 60 when first made. However, crosslinking of aldehyde groups on the rubber molecule causes the Mooney viscosity to rise in storage and in transit to the consumer, so that rubber received in a plant may well be between 75 and 100.

Because of its regular structure, natural rubber crystallizes when stretched, or when stored at temperatures below 20°C. The rate of crystallization varies with the temperature and also with the type of rubber, grades like pale crepe rubber (light colored rubbers often used in pharmaceutical products) freezing faster than smoked sheet. Since frozen rubber is almost rock-hard and practically impossible to mix, large users often have "hot rooms" to thaw frozen rubber they might receive. Temperatures are usually 50–70°C, and the bales are separated because of the low thermal conductivity of the rubber, 0.00032 cal/sec/cm^2/°C. Frozen rubber is not a common problem but it does occur (e.g., when rubber is stored or transported in unheated sheds or railway cars subject to low temperatures for days or even weeks).

In a typical natural rubber sample (Table 2.1), the acetone extract contains various sterols, esters, and fatty acids. Certain natural rubber antioxidants are also found in the acetone extract. The proteins and fatty acids are highly useful as vulcanizing activators.

B. Production

Commercially all natural rubber is derived from the species *Hevea brasiliensis*. This tree grows most readily in a band within 5° of the equator, in places where the annual rainfall exceeds 80 in./year, the temperature is 25–35°C, and low altitudes prevail. More than 80% of natural rubber comes from Southeast Asia:

Figure 2.4 Rubber plantation. (Courtesy of the Rubber Research Institute of Malaysia.)

production is about equal in Malaysia and Indonesia; Thailand accounts for much of the remainder in this area. *H. brasiliensis* is grown on estates and by small holders. By intensive plant breeding, tree productivity has been greatly increased. In the 40-year period 1930–1970 yield jumped from 1000 to more than 3000 lb/acre/year for the best stock.

Currently, clones in the Rubber Research Institute of Malaysia (RRIM) are yielding up to 6000 lb/acre/year in pilot scale plantings. A goodly portion of producing plantings are of older, less productive clones, and overall production yields are about 850 lb/acre/year. Ordinarily the trees are tapped when they are 5–7 years old and have a useful life of 25–35 years.

Figures 2.4–2.8 show typical scenes in the production of natural rubber. The plantation shown in Figure 2.4 is a well-kept one; the canopy of rubber trees is so thick that little grows underneath. In the single tree (Fig. 2.5), the white line at the bottom of the panel is latex being directed into the ceramic cup. The black dots (actually red) are spaced at regular tapping day intervals to record the number of days of tapping since that panel was started. Figure 2.6 shows the coagulating tanks at a rubber plant, with the central latex trough to channel latex to the individual tanks. Figure 2.7 presents a popular type of Malaysian technical specified rubber, and Figure 2.8 shows one of the crates normally used for overseas shipping.

Figure 2.5 Tapped rubber tree. (Courtesy of the Rubber Research Institute of Malaysia.)

The milky, rubber-bearing fluid in the tree is called latex and is obtained by a process called tapping. A cut, about 22° to the horizontal, is made into the bark of the tree cutting the latex vessels nearby. The latex exuding from the cut follows a vertical channel at the lower end of the cut into a small ceramic or glass cup. In the cutting process a small portion of the bark is excised. The latex flows for about 4 hr, and autocoagulation is prevented by placing a small amount of

Figure 2.6 Latex coagulating tanks in rubber factory. (Courtesy of the Rubber Research Institute of Malaysia.)

liquid ammonia in the cup. At the end of this prime flow period the latex is collected by the tapper and bulked in tanks. Usually 2 days later the autocoagulated material (cup lump) is removed from the cup and saved, a fresh cut is made beneath the previous one, ammonia is added, and the cycle repeated. Perhaps 80% of the flow is collected as liquid latex, up to 20% as cup lump, and very minor amounts collected as coagulum on the tree cut (tree lace) and at the foot of the tree (earth scrap). Latex fresh from the tree is called whole-field latex and will have 30–40% total solids by weight. The lower grades of natural rubber are prepared from cup lump, partially dried small holders' rubber, tree lace, and earth scrap, after appropriate cleaning.

The essence of dry natural rubber production is that whole-field latex, stabilized against coagulation by ammonia or other materials, is collected in bulk, screened to remove foreign matter, diluted, and the rubber coagulated with acid. The coagulum, a white spongy mass, is squeezed between contrarotating rollers to remove much of the water and serum and then dried to a moisture content of less than 1% by heat. Somewhat strangely, the squeezing operation uses a copious supply of iron-free water to wash the coagulum on the roll mills. The purpose is to wash out all traces of serum from the coagulum, the liquid that remains when the rubber is coagulated from the latex. Iron-free water is used to prevent oxidative deterioration of the rubber in storage. Variations in the feed mix, coagulation, and drying methods result in the different grades.

Figure 2.7 Polyethylene-wrapped Malaysian rubber bale. (Courtesy of the Rubber Research Institute of Malaysia.)

Despite acceptance of newer technically specified rubbers, a significant portion of the rubber used is called ribbed smoked sheet, usually abbreviated to RSS followed by the grade number. Ribbed smoked sheet is made from whole-field latex. It is diluted to about 15% solids and then coagulated with dilute formic acid. Coagulation usually takes place overnight. The coagulum is then passed through successive two-roll mills, which squeeze much of the water out. The last

Figure 2.8 Metric ton lot of technically specified rubber ready for overseas shipping.

pair of rollers has ribs that impart a ribbed appearance to the sheet and increase the surface area, to expedite drying. The sheets from the last mill are draped on poles, supported by a framework, which is wheeled into the smokehouse. Here the rubber is dried, very much like clothes on a line, for 2–4 days at 40–55°C. Losing its water, the white sheet turns a yellow-brown color in the smoke house and has a distinctive smoky smell. The rubber now has less than 1% moisture and is more fungus-resistant due to fungicides in the wood smoke.

For the lightest colored rubber products, a light-colored, premium priced, natural rubber called pale crepe may be used. This product calls for the use of clones having low yellow pigment content (mostly carotene) and higher resistance to discoloration by oxidation.

Technically specified natural rubbers were developed in the 1960s to make natural rubber more attractive in view of the competition from synthetic rubber. The production process is not much different from that for smoked sheet up to and including the coagulation stage. However, upon coagulation the sheet, after roll mills have pressed most of the water out of it and it is in the form of a continuous narrow blanket, is mechanically torn into crumbs. These are then dried in sieved-bottom trays by passing them through oil-fired air circulating tunnel driers at 100°C. Such a process can and does use such feedstock as cup lump and unsmoked partially dried sheet rubber. The dried crumb rubber is then compressed into 33⅓-kg bales, which have a standard size of 66 × 33 × 18 cm, wrapped in thin polyethylene film, and packed into crates containing 1 metric ton. Although installations for making technically specified block rubber require much more capital than traditional methods, they speed up the process, allow much more quality control, and package the product in a commercial way. The polyethylene film prevents adhesion between bales, keeps water out, and yet disperses easily when the rubber is mixed in the consumer's factory because of its low melting point.

Before the introduction of technically specified rubber, autocoagulated rubbers such as cup lump, tree lace, and earth scrap were washed, carefully blended, and milled till a relatively uniform product was obtained and dried. Some of these materials, called brown crepes, are still available, although a large and increasing percentage of feed material for these types now goes to the lower quality grades of technically specified rubbers.

C. Grades and Grading

All types of natural rubber that are not modified (such as oil-extended natural rubber) or technically specified rubbers (TSRs) are considered to be international grades. The current grades in this category are listed and specified in the booklet *International Standards of Quality and Packing for Natural Rubber Grades*, commonly called the Green Book from the color of its cover and obtainable in the United States from the Rubber Manufacturers Association. Also, master samples of the grades are kept at various organizations for reference and arbitration purposes.

Grade designations usually use color or how the rubber was made; typical grade descriptions are pale crepe #2, ribbed smoked sheet #1, and thin brown crepe #3. The main disadvantage of this system is that grading is done on visual aspects. Almost exclusively, the darker the rubber, the lower the grade. RSS #4, for example, is considerably darker than RSS #1. Other grading criteria, such as the presence or absence of rust, bubbles, mold, and wet spots, are subjective in nature. Perhaps the most valid assumption is that the darker, the rubber the more dirt it contains.

In the 1960s Malaysia led in developing a grading scheme that was more sophisticated and useful to consumers. A major criticism of natural rubber was the large and varying dirt content. The cornerstone of the new system was grading according to dirt content, measured in hundredths of 1%. For example, Standard Malaysian Rubber (SMR) #5 is a rubber whose dirt content does not exceed 0.05%. Dirt is considered to be the residue on a 45-μm sieve after a rubber sample has been dissolved in an appropriate solvent, washed through the sieve, and dried.

The specification has other parameters, including source material for the grade, ash and nitrogen content, volatile matter, plasticity retention index (PRI),

Table 2.2 ASTM Specifications for Technically Graded Natural Rubber

Property	Rubber grade[a]					
	Grade L	Grade CV[b]	Grade 5	Grade 10	Grade 20	Grade 50
Dirt retained on 45-μm sieve, % max	0.050	0.050	0.050	0.100	0.200	0.500
Ash, % max	0.60	0.60	0.60	0.75	1.00	1.50
Volatile matter, % max	0.80	0.80	0.80	0.80	0.80	0.80
Nitrogen, % max	0.60	0.60	0.60	0.60	0.60	0.60
Nitrogen, % min	0.25	0.25	0.25	0.25	0.25	0.25
Initial plasticity, min	30		30	30	30	30
Plasticity retention index, min	60	60	60	50	40	30
Color index, max	6.0					
Mooney viscosity		60 ± 5				

[a]Skim rubber is not permitted in any grade, and grades L, CV, and 5 must be produced from intentionally coagulated latex.
[b]Other Mooney ranges of grade CV are available: CV-50 ± 5 and CV-70 ± 5. When CV appears without a suffix, the 60 ± 5 form, as shown here, is assumed.

and initial plasticity [3]. Acceptance of these standards was vigorous, and other producing countries followed suit. Letter abbreviations identify the rubber source: SMR was rubber from Malaysia, SIR indicated Indonesian production, SSR Singapore, and so on.

Again following the Malaysian lead, standards organizations such as the American Society for Testing and Materials (ASTM) and the International Standards Organization (ISO) have evolved specifications of their own. Table 2.2 gives the ASTM specifications for technically graded natural rubber [4].

D. Modifications of Natural Rubber

Natural rubber has been modified in many ways since the establishment of a continuous supply of plantation rubber. Some of these modifications, such as chlorinated rubber, have passed their peak of acceptance as other materials supplanted them. Several others, however, can be very useful in compounding. In some instances the supply may be erratic, and care must be taken that an adequate and continuous supply will be available before adopting one for production runs.

Deproteinized rubber (DPNR) is a useful rubber when low water absorption is wanted, vulcanizates with low creep are needed, or more than ordinary reproducibility is required. Normally natural rubber has between 0.25 and 0.50%

nitrogen as protein; deproteinized rubber only about 0.07%. One tradeoff occurs here: since protein matter in the rubber accelerates cure, deproteinized rubber requires more acceleration.

Deproteinized rubber is made by treating natural rubber latex with a bio-enzyme, which hydrolyzes the proteins present into water-soluble forms. A protease like *Bacillus subtilis* is used at about 0.3 part per hundred of rubber by weight. The enzymolysis process may take a minimum of 24 hr. When complete, the latex is diluted to 3% total solids and coagulated by adding a mixture of phosphoric and sulfuric acids. The coagulated rubber is then pressed free of most of the water, crumbed, dried, and baled.

DPNR is viscosity stabilized at 60 Mooney. The heavy dilution of the latex helps reduce the dirt and ash content, and respective values of 0.006 and 0.06% are not unusual. Properly compounded, a black loaded stock (say 30 parts of SRF black) would give a stress relaxation rate of 2%/decade. Other vulcanizate properties would be a Shore A hardness of 50, elongation of 550%, and tensile strength of 3700 psi.

Oil-extended natural rubber (OENR) is also made in the Far East, especially Malaysia. Three ways have been used to make this type of rubber: (1) co-coagulation of latex with an oil emulsion, (2) Banbury mixing of the oil and rubber, and (3) allowing the rubber to absorb the oil in pans until almost all is absorbed, then milling to incorporate the remaining oil and produce a sheet. More recently, rubber and oil have been mixed using an extruder. Properties of the finished product do not differ significantly irrespective of the method used.

There is no strict limit on how much oil can be incorporated into natural rubber; it is a question of how oily a product can be handled. An arbitrary limit might be set as 65 parts of oil per 100 of rubber. Both aromatic and naphthenic oils are used.

Oil-extended natural rubber masterbatches are used where high black and oil contents are wanted in a natural rubber stock or blends with natural rubber. Incorporation of large amounts of oil is easier if some portion of the oil is already in the rubber. Another advantage is that the masterbatches are freeze-resistant. Oil-extended natural rubber masterbatch has not been used much in North America but has enjoyed some popularity in Eastern Europe and Japan.

Latex is not the only form of liquid rubber; depolymerized rubber is also available in certain areas. In the United States it is made by Hardman Inc. of Belleville, NJ, under the tradename DPR. Production details are proprietary. The product looks somewhat like molasses and comes in two grades, a low viscosity grade (45,000–75,000 CP/sec at 100°C) and a high viscosity grade (175,000–40,000 CP). Viscosity measurements are made in a Brookfield viscometer. These rubbers can be compounded in a split batch compound to give self-curing compounds as well as by using a conventional cure system vulcanizable by heat. These rubbers fill a niche for applications in which conventionally prepared rubbers would be unsatisfactory. For example, they make excellent flexible molds to release waxes, gypsums, and ceramics easily with excellent surface reproduction and dimensional stability. Other uses include binders in grinding wheels, dispersants, and automotive sealants. DPR compounds are not recommended for outdoor exposure.

A comparatively new modification of natural rubber produced in commercial quantities is epoxidized natural rubber (ENR). The rubber molecule is partially epoxidized, with the epoxy groups randomly distributed along the molecular chain. Commercially two grades were developed, identified by the extent of the modification: 25 mol % epoxidized and 50 mol % epoxidized. With epoxidation, a segment of the molecular chain looks like this:

These products, formed by the addition of oxidizing agents on latex, were developed by the Malaysian Rubber Producers Research Association (MRPRA) in conjunction with the Rubber Research Institute of Malaysia. The developers used the designations ENR 25 and ENR 50; commercial brands may carry other names. Kumpulan Guthrie Berhad in Malaysia, for example, have described the properties of their Epoxyprene brands in a series of studies [5].

The properties of these rubbers are so unique they might be considered to be new elastomers. Although the Mooney viscosity at 70–100 parallels much natural rubber, the specific gravity increases significantly, as does the glass transition temperature, as epoxidation is increased. For the 25 and 50% epoxidized materials, one commercial producer lists specific gravities of 0.97 and 1.03 and glass transition temperatures of −45 and −20°C, respectively.

Of interest to compounders are the following advantages the epoxidized polymers have over natural rubber:

1. Improved oil resistance—ENR 50 is similar to chloroprene rubber.
2. Low gas permeability—ENR is similar to butyl rubber.
3. Because of the polarity induced by the epoxidation, these rubbers are compatible with polyvinyl chloride, which suggests their use in adhesives.
4. With epoxidized rubber and precipitated silica as the reinforcer, of the reinforcement properties obtained are very similar to those of natural rubber and carbon black. This is done without the use of expensive coupling agents.

In processing and compounding, epoxidized rubbers differ from conventional rubbers. Little if any premastication is required. It is easy to overmix the polymer. Vulcanization systems commonly used with unsaturated polymers can be selected, but semiefficient or efficient systems give better aging. Thermal degradation of epoxidized rubber occurs differently than natural rubber and it is customary to add 2–5 parts of calcium stearate to combat it. Protective systems against ozone attack are generally higher than with unmodified natural rubber, and levels of 3.5–6.0 parts per hundred of rubber are not uncommon.

E. Some Compounding Considerations

Briefly, the most significant properties from the compounder's point of view are dirt, ash, nitrogen content, volatile matter, plasticity retention index, and Mooney viscosity. It is known that the higher the dirt content, the greater the chances of failure of rubber parts from fatigue. High ash content may mean rubber contaminated with sand or even, very rarely, adulterated. Volatile matter should be low because it is a measure of the water present, and wet spots are hard to disperse in mixing. Nitrogen in amounts over the specification limit could indicate the presence of skim rubber: this by-product rubber from concentrated latex manufacture is high in protein, which makes its rubber very fast-curing. The plasticity retention index is a measure of the oxidation resistance of natural rubber at a specified temperature. The higher the value, the more resistant the rubber. The higher the viscosity, the more breakdown time is required. In some cases premium priced viscosity-stabilized rubber may be cost efficient.

Some general considerations regarding the use of natural rubber should be kept in mind by the compounder. Although NR prices have been relatively stable during the last few years, natural has had a history of wide price swings, so up-to-date information is a must. Large consumers can protect themselves after a fashion by buying futures on the London market. Natural rubber is unique among elastomers in being traded in this way. There is little chance that there will be a shortage of natural rubber in the foreseeable future. A sharp increase in demand will encourage small holders to tap regularly and more extensively. Latex yield promoters such as ethylene producing Ethrel TY can increase flow by 25% or more, the response is different with different clones.

For efficient, trouble-free use of NR, the compounder should observe the following guidelines.

1. Stock only the types needed; four grades should be ample.
2. If volume permits, try to buy from a single estate. A list of Malaysian growers, packers, and shippers is available from the Malaysian Rubber Bureau USA.
3. Inspect shipments on arrival for pieces of paper, crate wood embedded in the rubber, or "knuckles" (small cream-colored areas, which are high in moisture). If such defects are present, take the matter up with the supplier.
4. Protect the rubber from sunlight, cold weather, and water.

Some trends have continued in technically specified and latex rubbers in the last 5 years. For example, there is a diminishing supply and demand for pale crepe rubber, which is being replaced in many instances by rubbers like SMR L and SIR L. At the other end of the color and quality scale, technically specified rubber 50 is little used.

IV. SYNTHETIC POLYISOPRENE

It has been known for decades that natural rubber is a polymer of isoprene. The importance of natural rubber in World War II was a strong inducement for U.S. scientists to develop a synthetic polyisoprene, which would lessen the country's dependence on overseas supplies. Early attempts to duplicate natural rubber were of limited success because of the heterogeneity of the molecular configurations. The development of stereospecific catalysts in the late 1940s and the 1950s changed this. These agents have the ability to make the monomer units join in definite selected arrangements. The first commercial synthetic polyisoprene was introduced by the Shell Chemical Company in 1960. This was followed by one from Goodyear in 1962, and three years later Goodrich Gulf introduced their version. Today synthetic polyisoprene is made worldwide by Japan, Italy, Romania, and other countries. Technically speaking, synthetic polyisoprene is a close match for natural rubber. Because of its cost, it has not supplanted natural to any significant extent.

Although techniques for manufacturing synthetic polyisoprene are largely proprietary, it is known to be made by a solution rather than an emulsion process. A typical method might be charging high purity isoprene, solvent, and catalyst to a reactor. Ziegler–Natta catalysts are used, and the choice of such catalysts has a significant effect on properties. When the targeted conversion of monomer is reached, the addition of a short stop deactivates the catalyst, and the polymer is stabilized against oxidation by adding an antioxidant. The polymer can be recovered from the rubber cement by steam stripping the solvent, converted to crumb form by hot water, then dried and baled.

Synthetic polyisoprene and natural rubber are seen to be close in many properties when it is realized that more variation can be expected in the natural product. Mooney viscosity (ML 4 @ 100°C) for synthetic polyisoprene is about 75–95, whereas the envelope for natural is around 50–90. The expression in parentheses is the shortened form indicating that the viscosity involved is Mooney; L stands for the large rotor that was used, and the 4-minute value at 100°C is the one reported. Number-average molecular weight for a typical synthetic polyisoprene is 350,000–400,000; that for natural rubber is 300,000–500,000. Glass transition temperature (the temperature at which the polymer goes from a stiff to a glassy state) is about -70°C; for a typical synthetic polyisoprene, about -72°C. Synthetic polyisoprenes are water-white; this is matched by only one or two grades of pale crepe natural rubber.

Some vulcanizate properties of synthetic polyisoprenes can at times justify their usually higher price. Contamination of synthetic polyisoprene is rare, whereas some dirt is usually present in natural rubber—in fact, the grading system is based on dirt content, as noted earlier. Although this property is difficult to assess, uniformity of synthetic polyisoprenes is probably higher than that for natural rubber. This would be a consideration in, say, producing large volumes of an extruded item to be cured at high speed by passing through a molten salt bath.

The compounding and processing of synthetic polyisoprenes is similar to that for natural. In compounding it must be remembered that tree rubber has naturally occurring fatty acids, accelerators, and antioxidants, which synthetic polyisoprene lacks. Accordingly, at least 2 parts of stearic acid should be added as an activator and perhaps 10% more accelerator and antioxidant when compared to a natural rubber formula.

In processing it is very easy to overmix synthetic polyisoprene. Often preliminary breakdown can be omitted and one-step mixing used. If feasible, open mill mixing should be adopted and the antioxidant added to the rubber as soon as it is banded on the mill. This will help prevent chain scission of the lower molecular weight portion due to oxidation. Remilling for extrusion or calendering should be kept to a minimum. Synthetic polyisoprene products range from bottle nipples and elastic thread to sewer pipe gaskets and blowout preventers.

REFERENCES

1. *1989 Annual Book of ASTM Standards*, Vol. 09.01, p. 293.
2. M. Morgan, in *Science and Technology of Rubber*, F. R. Eirich, ed., Academic Press, New York, 1978, p. 25.
3. *1989 Annual Book of ASTM Standards*, Vol. 09.01, p. 529.
4. *1988 Annual Book of ASTM Standards*, Vol. 09.01, p. 408.
5. Documents ENR-01-1 through ENR-07-11, published by Kumpulan Guthrie Berhad, Wisma Guthrie, 21 Jalan Gelanggang, Damansara Heights, 50780 Kuala Lumpur, Malaysia.

3

Elastomers: Styrene-Butadiene Rubber

I. SCOPE

This chapter reviews styrene-butadiene rubber (SBR), by far the world's fore-most synthetic rubber and easily available world-wide. This is because prac-tically all automotive tires use SBR, especially in the tread. Under normal driving conditions, in the countries having the highest car ownership, SBR treads have a wear advantage over natural rubber treads.

SBR was developed by German scientists in the 1930s to lessen that country's dependence on foreign sources for commodities. With the outbreak of World War II in 1939 and disruption of natural rubber shipments in 1941, intense and successful efforts were made in the United States and Canada to create SBR production facilities to supply the Allies' military and civilian requirements. A concurrent problem was to improve on the German product, which was difficult to process. In the decade after the war, all U.S. government-built and -owned plants were sold to private industry. In Canada the facilities remained a Crown corporation for over 40 years but were finally privatized.

A. Properties

Styrene-butadiene rubber is a copolymer of styrene and butadiene normally containing about 23.5% bound styrene. It is made both by emulsion and in solution polymerization. The repeating unit in SBR is shown in Figure 3.1. Because of the double bond present in butadiene, that portion of the molecule can

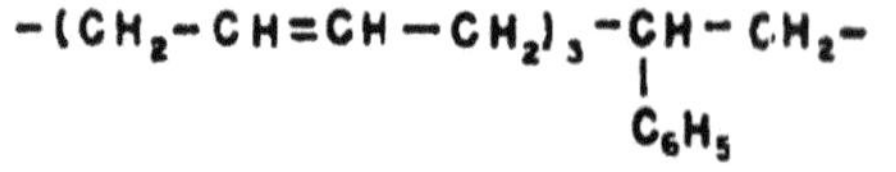

$$-(CH_2-CH=CH-CH_2)_3-CH-CH_2-$$
$$| $$
$$C_6H_5$$

Figure 3.1 Repetitive unit in SBR rubber.

be present in three forms: *cis*-1,4, *trans*-1,4, and vinyl-1,2. Ordinarily the proportions are about 18% cis, 65% trans, and 17% vinyl. Solution SBR has about the same amount of styrene as the emulsion product, but there is a higher *cis*-butadiene content, lower *trans*-butadiene, and lower vinyl. Mooney viscosity of unpigmented SBR made by the emulsion process ranges from 23 to 128, with the most popular grades being between 44 and 55. This viscosity measurement is the ML 4 (100°C) viscosity as described in ASTM Test Method D1646 [1]. Density of SBR is 0.94. The T_g of SBR is higher than natural rubber, a common grade of SBR, S 1500, having a T_g of about $-53°C$, whereas natural rubber is about $-70°C$. T_g is of course the glass transition temperature at which the rubber goes from a rubbery to a brittle glassy state.

B. Manufacture

The great majority of SBR is made by emulsion polymerization. The two monomers in styrene-butadiene rubber are of course styrene ($C_6H_5CH=CH_2$) and butadiene ($CH_2=CH—CH=CH_2$). Styrene is prepared by catalytically dehydrogenating ethyl benzene in the vapor phase. The source of the ethyl benzene is usually the alkylation of benzene with ethylene. Butadiene is made by the dehydrogenation of *n*-butane or of *n*-butene by the Houdry process. Coproduct butadiene is recovered from the cracking of hydrocarbons such as gas oil, and naphtha fractions. Solvents such as furfural, dimethylacetamide, or dimethylformamide are used to extract butadiene from crude butadiene.

The formulator of an emulsion SBR recipe has a wide variety of "handles" available. These can include the relative amounts of styrene and butadiene used, the temperature of polymerization, the percentage conversion, and the antioxidant choice. Apart from manufacturing efficiency, variations in polymerizing conditions can cause changes in the processing and vulcanizing characteristics of the rubbers produced, and the compounder should be aware of these.

One of the most important variables is the monomer ratio in the copolymer. This shows up dramatically in the change in the glass transition temperature. Polystyrene has a glass transition temperature of about 90°C, polybutadiene's glass transition temperature is about $-90°C$. As the weight fraction of styrene in the copolymer increases, there is a regular increase in T_g. At 0.25 weight fraction the glass transition point is about $-65°C$, at 0.5 about $-55°C$, and at 0.75 about 0°C. The influence of increasing styrene ratio on the properties of the vulcanized rubber compounds is less clear. Storey and Williams [2] investigated this phenomenon and found that as the styrene content increased, modulus increased, but resilience decreased. Tensile strength increased as the styrene content increased to 0.5 weight fraction, and then decreased. As styrene content is increased, the polymer loses its rubberiness and becomes more like a plastic in processing. For example, extrusion speeds increase, and there is reduced die swelling and mill shrinkage. The last two effects are defined as follows.

1. When a rubber compound is extruded through an orifice (e.g., a circular die), there usually is an increase in diameter from the size of the die. This ratio, known as *die swell*, usually is expressed as the percentage increase in cross-sectional area.
2. Similarly, after passage through a roll mill, a blanket of rubber or rubber compounds will shrink with time from its original lateral dimensions and become thicker. This is known as *mill shrinkage*.

Materials with high styrene contents (say, 80:20 to 90:10 ratios) are produced by emulsion polymerization and identified as high styrene resins. These are resins, not rubbers, and they are used as organic reinforcers to stiffen rubber compounds. High styrene latices can be blended easily with other latices to give stronger films.

Probably of equal importance in its effect on processability and compounding properties of the copolymer is the temperature at which the reaction is carried out. It was realized early that polymerizations for rubbery materials resulted in better products if carried out at normal rather than elevated temperatures, but reaction rates were slow. The full value of low temperature polymerization was achieved shortly after World War II by the use of redox initiation systems, which allowed satisfactory conversion rates at such low temperatures as 5°C. The name *redox* was coined because the process involves the interaction of a reducing agent and an oxidizing agent.

One of the properties affected by the polymerization temperature is the tensile strength of the copolymer. Over the temperature range 65–5°C, the tensile strength increases almost linearly as the temperature drops. Whereas a polymer made at 65°C with a 50-part loading of carbon black might have a tensile strength of 2600 psi, rubber with the same amount of black but made at 5°C might show 3900 psi, a 50% increase. Coupled with the increased tensile strength are better abrasion resistance and improved resistance to flex cracking. These qualities are important in tires, so the use of cold rubber, that made at 5°C, is understandable. The benefits as usual are not without tradeoffs. Compared to emulsion SBR made at 55°C, cold rubber is harder to process, and extrusions and calenderings tend to be rougher.

Two other polymerization conditions have effects on polymer properties that are worth noting. Conversion of monomer to polymer is usually stopped before the 70% conversion point is reached. The major reason for this is that as the percentage conversion goes past this level, the amount of highly branched and crosslinked polymer rapidly increases. The presence of this polymer gel makes the rubber hard to process. It shows a nerviness or toughness on the mill that makes plastication difficult.

Another undesirable feature that occurs as conversion approaches 70% and higher is the growing heterogeneity of the product. Over 60%, conversion the molecules that form are richer in styrene, hence less compatible with earlier formed polymer. Finally, the amount of modifier has a great effect on the polymer produced. If considerable excess is used, the polymer gel content may be low; but in the worst case, the product is a liquid rather than a rubber.

To avoid duplication of research efforts in World War II and later until the synthetic rubber facilities were sold to private owners, there was a free exchange of information among producing plants. As a result of this cooperation, a common formula for making SBR at 50°C was developed known as the "mutual" recipe. It is shown in Table 3.1. Later a recipe known as the "custom" recipe was derived for making SBR at 5°C (Table 3.2).

Most emulsion SBR is made by a continuous process. In the recipe for hot rubber, water and the fatty acid soap provide the aqueous phase for the emulsification of the monomers, lauryl mercaptan is the modifier regulating molecular weight, and potassium persulfate initiates the reaction. The cold rubber formula is somewhat more complex. The first five ingredients serve the same purpose as their counterparts in the hot rubber formula. Rosin acid soap is chosen because it has greater solubility in cold water than conventional soaps. The redox initiation system is the cumene hydroperoxide and the ferrous sulfate heptahydrate couple. Here the hydroperoxide is the oxidizer and the iron salt the reducing agent. The purpose of he pyrophosphate is to form a complex with the ferrous salt, so iron(II) ions are present at the correct low concentration.

During initiation, the iron(III) ions that are produced are reduced back to iron(II) by dextrose, a reducing sugar. Using inorganic electrolytes like po-

Table 3.1 "Mutual" SBR[a] Recipe

Component	Parts by weight
Styrene	25
Butadiene	75
Water	180
Fatty acid soap	5
Lauryl mercaptan	0.50
Potassium persulfate	0.30
	285.80

[a]Early publications used GR-S (government rubber—styrene) instead of SBR. The term GR-S is now obsolete.

Table 3.2 "Custom" Recipe for 5°C SBR

Component	Parts by weight
Styrene	28
Butadiene	72
Water	180
Potassium soap of disproportionated rosin acid	4.7
Mixed tertiary mercaptans	0.24
Cumene hydroperoxide	0.10
Dextrose	1.00
Ferrous sulfate heptahydrate	0.14
Potassium pyrophosphate	0.177
Potassium chloride	0.50
Potassium hydroxide	0.10
	287

tassium chloride ensures that the reaction medium stays fluid during polymerization.

Undoubtedly in the decades after the polymer plants were put in private hands, changes have been made in these recipes, and much current production is made by proprietary methods. However, the above illustrates typical ingredients in the polymerization formulas.

In continuous emulsion polymerization of styrene and butadiene to make SBR, the various ingredients are charged continuously to reactors in series. These reactors are fitted with agitators and temperature control facilities. Rates of flow are adjusted so that the efflux from the last reactor in the series is at the desired conversion. This might be 70% conversion for rubber made at 50°C; cold rubber might be stopped at 60–65% conversion. After being discharged from the reactors, the product is treated with a short-stop solution—that is, a compound added to terminate polymerization activity. Hydroquinone has been used with hot rubber, sodium polysulfide for cold rubber.

The product, now called latex, is transferred to a blowdown tank. The latex is steam heated, and part of the unreacted butadiene is flashed off; the residual butadiene is removed by vacuum. The recovered butadiene is recycled. Again using steam, unreacted styrene in the latex is distilled off and also recycled. The final common step in SBR manufacture is the addition of an antioxidant dispersion. Since mechanical stresses in the baling operation are considered to be capable of initiating oxidative attack, the rubber must be protected against deterioration by oxygen.

A common method of coagulating the rubber from the latex has been to add brine, which causes a concentration of rubber particles at the top of the solution, followed by the addition of acid, which coagulates the rubber hydrocarbon into a floc. This is filtered out in drum filters and washed. Other methods of coagulating the rubber from the latex are the use of glue/acid and alum. The floc is usually dried on belt dryers. Some water can be removed prior to belt drying by passing the coagulum through dewatering extruders, which may bring the water content down to about 15%. After drying, the polymer is checked for metallic contamination, pressed into bales, and wrapped in polyethylene film; then, with further protection, it is ready for shipment. Variations in this process can occur with oil emulsions and/or carbon black dispersions being added to the latex after the short-stop addition. In such ways a wide variety of masterbatches is provided for the compounder.

Very popular forms of modified SBR are the masterbatches made with oil, with black, or with oil and black. Dealing first with masterbatches containing rubber and oil alone, oil is added only to cold polymerized SBR. When used at relatively low amounts (say 5–12 parts), oil is added to improve processibility and to cheapen the product at high loadings. Only cold rubber is used because it can be made with Mooney values in the 110–135 range, so by co-coagulation of such latex with the proper amount of oil a processible polymer is produced at low cost. Oil masterbatch is usually made by a continuous process in which an oil emulsion and the latex are mixed thoroughly in the desired proportions and then coagulated, filtered, dried, and baled.

The oils used in making these masterbatches are petroleum oils; others would be too costly. With rare exceptions, petroleum oils classed as highly aromatic (HI-AR) or naphthenic (NAPH) are used. Paraffinic oils are not highly soluble in SBR and have a tendency to bloom or sweat out. Masterbatches containing aromatic oils are considered to be staining; the naphthenics, nonstaining. Cold rubber oil masterbatch usually has 37.5 parts of oil per 100 parts of rubber (phr). Cold rubber oil black masterbatch has 15–60 phr. Carbon black loadings in the masterbatch will vary from 75 to 100, the most common levels being 75 and 82.5 phr.

Before going on, it is useful to describe the classification scheme devised by the International Institute of Synthetic Rubber Producers (IISRP), which has been generally accepted. All hot or cold polymerized styrene butadiene rubbers with or without oil and/or carbon black are given code numbers with six numbered series.

The series numbers are as follows:

1000   Hot nonpigmented rubbers
1500   Cold nonpigmented rubbers
1600   Cold black masterbatch with 14 or less parts of oil per 100 parts
         SBR
1700   Cold oil masterbatch

1800   Cold oil black masterbatch with more than 14 parts of oil per 100 parts SBR

1700   Emulsion resin rubber masterbatches

Somewhat arbitrarily, it is considered that up to 14 parts of oil might be used with SBR of average molecular weight in compounding but more oil would be necessary with high molecular weight rubber to keep processibility. This practice of using more than 14 parts of oil in the polymer is called oil extension.

Starting in the 1920s many attempts were made to produce natural rubber carbon black masterbatches by blending latex with a carbon black dispersion and co-coagulating. These attempts were largely unsuccessful. The reinforcing blacks at that time were limited to channel-type blacks, which are absorptive and sometimes negated the effect of the natural antioxidants present with the latex. It is also difficult to make such a masterbatch that will not harden on storage.

Synthetic rubber masterbatches, especially SBR masterbatches, have been more successful. In this process a water dispersion is made of the black by high speed stirring with the help of a dispersing agent, perhaps a lignin sulfonate. In such dispersions the aggregates are similar in size to the rubber particles in the latex. Care has to be taken that the dispersing agent for the black and the stabilizer for the latex are compatible. The coagulation step is next, and it is important that there be good dispersion of carbon black in the crumb, which is filtered out, dried, and baled. By adding an oil emulsion as well before coagulating, a three-part masterbatch is produced (i.e., rubber, oil, and black).

Masterbatches such as these have several advantages for the compounder. If the masterbatch requires little modification in the plant—that is, if the rubber, black, and oil are in the right proportions—it may well be cheaper to buy masterbatch than to do the mixing in house. Certainly in the earlier days of SBR masterbatches, when the product was priced at little over the cost of the separate ingredients, the compound was getting the mixing almost free. In the plant there are economies in storage, and in mixing processes less energy and less dust. In many cases dispersion of the finished compound is improved over mixing in the plant. This is not always the case. Very fine blacks like Super Abrasion Furnace seem as hard to disperse in polymer plant masterbatching, if not more so, than mixing on roll mills or in internal mixers. A disadvantage of SBR masterbatches is that there may be differences in cure rate whose causes are difficult to identify.

One convention generally observed in compounding with these masterbatches should be noted. At one time there were heated discussions as to whether the added oil should be treated as rubber in determining curative dosages, etc. This question has now been resolved. The formulas are written with the amount of masterbatch that contains 100 parts of rubber, and the added ingredients then are those normally used with 100 parts of rubber. For example, if masterbatch 1815 is being used, the first line in the compound would be:

    1815 MB   225.0

as this masterbatch contains 75 parts of high abrasion furnace (HAF) black and 50 parts of naphthenic oil.

C. Grades and Grading

Reference has already been made to the classification scheme for SBR rubbers developed by the IISRP. Although the list is not a permanent one (old code numbers are no longer produced and new ones are added), a recent listing had over 80 different rubbers. Many of these coded rubbers are made by more than one producer. The IISRP listing gives properties of interest to the compounder for each code. These include the percentage of bound styrene, the Mooney viscosity, the types of soap used in the recipe, the coagulant, whether the product stains, the specific gravity, and the amount and type of oil and black added if the rubber is a masterbatch. A common cold rubber is 1500 in the IIRSP listing. This rubber has 23.5% bound styrene; the Mooney viscosity is 52, and a rosin acid soap is used as the emulsifier. Acid or salt/acid is the coagulation method used. One of the most popular oil rubber masterbatches is 1712, for which the same kind of information is given, along with the advice that 37.5 parts of an aromatic oil has been added. As noted above, cold polymerized carbon black–oil–rubber masterbatch like 1815 contains 75 ph HAF plus 50 pphr of naphthenic oil. Incidentally, high black–high oil masterbatches like this have been used because of their cheapness as a replacement for reclaimed rubber. They are also used frequently in competitive grades of retread rubber.

Besides divulging the properties described above, SBR producers routinely give the results of their products when tested in evaluation formulas. Table 3.3 is the specific formula being used by the ASTM and the ISO to test rubber. Table 3.4 gives the formula developed by the ASTM where the product contains carbon black or carbon black and oil. In these procedures the mixed compounds are vulcanized for 25, 35, and 50 min at 293°F. From the stress, strain, and hardness values on such samples the compounder can determine whether the rubber cures fast enough for his purpose and will impart the necessary strength, stiffness, or hardness.

Earlier the term *masterbatch* was used. The ASTM has defined this term as follows:

> A homogeneous mixture of rubber and one or more materials in known proportions for use as a raw material in the preparation of the final compounds. Note—Masterbatches are used to facilitate processing or enhance the properties of the final product or both.

Table 3.3 Evaluation Formula for SBR

Component	Quantity (parts by mass)
SBR or SBR masterbatch	100.00
Zinc oxide	3.00
Sulfur	1.75
Stearic acid	1.00
Oil furnace black	50.00
TBBS[a]	1.00
	156.75

[a]TBBS = *N-tert*-Butyl-2-benzothiazolesulfenamide.
Note: Standard reference materials (SRM) obtained from the National Bureau of Standards (NBS) are used with the exception of the rubber under test. NBS became National Institute of Standards and Technology in 1988.

Table 3.4 Evaluation Formula for SBR Mixed with Carbon Black or Carbon Black and Oil

Component	Quantity (parts by mass)
Masterbatch	$100 + X^A + Y^B$
Zinc oxide	3.00
Sulfur	1.75
Stearic acid	1.50
TBBS[a]	1.25
	$107.50 + X + Y$
	where X^A and Y^B are parts of carbon black and oil, respectively, per 100 parts of base polymer

[a]TBBS = *N-tert*-Butyl-2-benzothiazolesulfenamide.
Note: Standard reference materials (SRM) obtained from the National Bureau of Standards (NBS) are used with the exception of the masterbatch under test. NBS became National Institute of Standards and Technology in 1988.

The advantages of using an SBR masterbatch have been noted. However, rubber goods manufacturers may not only buy masterbatches, they frequently make their own. An example might be the use of an accelerator masterbatch. Accelerators, described later, are often used in minute quantities in a rubber batch. If care is not taken, a significant portion may be lost by dusting, by lodging under the guides of a roll mill, or otherwise. If, however, a masterbatch of, say, 20 parts accelerator and 80 parts rubber is first made and a proper aliquot is used, loss is smaller and the accelerator is dispersed more completely.

D. Some Compounding Considerations

Compared to natural rubber, SBR is somewhat harder to break down in mixing for easier incorporation of other components of the compound. It does not freeze as natural does; instead, it hardens with age. With less unsaturation it is slower curing than natural, but it shows fewer symptoms of overcure. It is difficult to make pure gum stocks from SBR. These compounds, which consist usually of rubber, vulcanizing agents, and age resisters, are exemplified by rubber bands. SBR does have better abrasion resistance than natural in such items as passenger tire treads, although in winter weather, natural is superior. Either natural or SBR can be used for many products, ard the choice may well depend on the relative prices of these two rubbers.

If service conditions for a product indicate that SBR would be a rational choice, the compounder has a wider range of types and grades that can be obtained with natural. Questions to be answered include whether to use hot or cold polymer, what antioxidant system to use, whether staining is a disadvantage, and whether the curing rate is adequate. A further use of masterbatch may be cited: if the plant's output is mainly light-colored or white goods—sports equipment, perhaps—then contanination of mixed batches by loose carbon black might be avoided by purchasing black masterbatch for occasional black mixes.

E. Solution SBR

Although solution-polymerized SBR has been available for some years, only comparatively recently has much interest been shown in its use. The increased interest is due to demands by automotive manufacturers for tires with low rolling resistance, high wet and dry traction, and excellent wear. These are conflicting demands, but solution SBR offers promise in meeting them. Much of the present capacity for solution polymers is in developing countries, where tariff protection encourages surplus production. New SBR plants are usually solution not emulsion facilities. With this wide availability, solution SBR should not be of interest to tire compounders alone. Perhaps due to the utility of this polymer in tires and the competitiveness in that market, relatively little has been published on its use.

Solution SBR is available as dry or oil-extended rubber. The dry rubbers are practically water-white and approximately 98% by weight polymer compared to the 92–94% typical of emulsion types. Specific gravity is usually 0.93–0.95. Bound styrene can be from 10 to 60%, but most grades are at 18 or 23.5–25.0%. Cis content is usually 34–36%; vinyl tends to be less than 15%, very frequently 8–10%. Although some special grades have viscosity levels, ML $1+4$ (100°C), as low as 13 and as high as 90, most grades marketed range from 45 to 50. If the extender oil is naphthenic, the polymer is usually nonstaining or slightly staining; if aromatic, it is staining.

With the wide range of grades in both emulsions and solution-polymerized rubbers, comparison of their processing and vulcanizate properties can be difficult. If solution polymer 1216 (regular IISRP number), non-oil-extended with 50 parts HAF black loading, is compared at the same loading with S-1500, the former normally would show faster curing (up to ⅓ less time) and 300–500 psi lower tensile strength. Hysteresis is usually lower as well.

Inasmuch as the thrust of research activity with SBR seems to be on solution polymers, some papers on this subject are briefly noted. Although the literature focus is on tire treads, the information gleaned may be useful in compounding other rubber products.

An early informative paper on the use of solution SBRs was that by Moore and Day [3]. Some of their conclusions follow: when comparing emulsion versus solution SBRs having the same bound styrene level, the solution SBR provides lower rolling resistance, improved dry slide traction, and superior wear resistance; the emulsion SBR provides better wet slide traction. If one compares emulsion and solution SBRs based on the same Young's modulus index (YMI), the solution SBR is found to be better for rolling resistance, dry slide, and wet peak traction, while the polymers are equal for wear. Comparing emulsion and solution polymers based on YMI appears to give the potential for breaking out of one of the most difficult compromises facing the tire compounder. That is, both wet traction and rolling resistance now can be improved with no sacrifice in wear.

In tire technology the term tan δ (tan delta) is frequently used in describing a compound's behavior under forced vibration. More specifically it is a term used when the motion is sinusoidal. It is defined in the ASTM Standard Practice for Rubber Properties in Forced Vibration as follows:

loss factor (loss tangent, tan δ, arc G^x) − the ratio between loss modulus and storage modulus tan δ = G''/G'

The loss modulus (viscous modulus, hysteretic modulus, imaginary modulus, out of phase modulus), G'', is the component of applied shear stress that is 90° out of phase with the shear strain, divided by the strain. The storage modulus (elastic modulus, real modulus, in-phase modulus), G', is the component of applied shear stress that is in phase with the shear strain, divided by the strain.

The value of tan δ does change with temperature. At 40°C, tan δ correlates with rolling (resistance) loss, with rolling loss increasing as tan δ increases.

A paper by Aggarwal et al. [4] deals with experimental solution polymers in tires. They point out that rolling resistance and wet traction are related to low tan δ and T_g, respectively. Conventional synthetic rubbers have high T_g and high tan δ. Tan δ as determined by an instrumented pendulum skid tester showed excellent correlation with wet skid coefficient of friction of an actual tire test. The authors prepared several solution polymers including SBR by different anionic polymerization techniques. Unlike conventional rubbers in which tan δ and T_g increase in parallel from cis BR to emulsion SBR 23.5 styrene, the new rubbers have significantly higher T_g values coupled with low tan δ. Accordingly these new rubber compositions have superior low rolling resistance and high wet traction, compared with conventional rubber tires.

Futamura [5] examined the SBR polymer properties that had a large effect on tire traction. The author's conclusions are too lengthy to be included here. One finding was that when soap residue is removed from emulsion SBR by extraction with toluene/ethanol, the resulting product is very similar to solution SBR. For example, cure rate was increased, hysteresis and tear strength were reduced, and Mooney viscosity was increased. The author however points out that the residues contribute positively to the better processibility and higher strength of emulsion SBR.

REFERENCES

1. D1646-90, *1991 Annual Book of ASTM Standards*, Vol. 09.01, p. 308.
2. Storey, E. B., and Williams, H. L., A Study of Styrene-Butadiene Copolymers, *Rubber Age*, *68*, 571–577 (1951).
3. Moore, D. G., and Day, G. L., Comparison of Emulsion vs. Solution SBR on Tire Performance, Firestone Synthetic Rubber and Latex Co., Akron, OH., (1985).
4. Aggarwal, S. L., Hargis, I. G., Livigni, R. A., Fabris, H. J., and Marker, L. F., Structure and Properties of Tire Rubbers Prepared by Anionic Polymerization, in *Advances in Elastomers and Rubber Elasticity*, Lal, J. and Mark, J. E., Eds., Plenum Press, New York, (1986).
5. Futamura, S., Central Research Laboratories, Firestone Tire and Rubber Company, Winter Meeting Technical Papers 1988, Akron Rubber Group, 1988.

4

Elastomers: Polybutadiene and Chloroprene Rubber

I. SCOPE

This chapter reviews two commonly used rubbers, polybutadiene and neoprene. Polybutadiene is almost always used in blends with other rubbers, a major reason being the difficulty of mixing it alone. Among other useful qualities it has excellent abrasion resistance, which has won it a preferred place in tire compounding, especially passenger tire treads. Neoprene, developed as a special-purpose rubber to give good resistance to petroleum hydrocarbons (an early use was in gasoline pump hose), remains a leading synthetic rubber despite competition from other synthetics.

As a guide to selecting the most appropriate polymers for a given application, a largely subjective listing of the properties of many of the polymers discussed here is given in a later chapter (see Table 6.1). Two rubbers are not included in that list: polysulfide rubbers and reclaimed rubber. Polysulfide rubbers are used only in very specific applications and normally would not be considered to be optional rubbers. Reclaimed rubber is not included because of the dearth of up-to-date data on many properties used in Table 6.1. Allowing for these omissions, it should be realized that many compounds fall into a gray area where more than one polymer may be quite adequate, in which case the lower cost polymer is chosen.

As mentioned above, polybutadiene is often used in blends. It may not be realized that a major portion of rubber stocks mixed in this country are blends.

This is because the tire industry regularly prefers blends for its large-volume compounds, such as tread and sidewall stocks. Blends do not always involve two polymers. In many cases, such as white sidewall, three polymers may be used. Although it does make compounding more complex, nontire compounders should always consider going this route as well.

In an early unpublished paper, the author reviewed data on blends of BR, SBR, and NR. Some effects appeared that would not have been inferred from a calculation estimating properties on the relative proportion of the polymers in the blend. For example, if a 100% polymer A compound had a tensile strength of 4000 psi and a 100% B compound 2800 psi, it might be estimated that a 50:50 blend would give an average tensi e strength of 3400 psi. In the author's studies, such mixture averages were not usual. For example, a 50:50 blend of an NR stock and a high cis BR stock has much higher tensile strength and abrasion resistance than would be expected considering the values of the single-polymer stocks. This is given as a precautionary note; use only estimates based on single-polymer properties as a preliminary guide in setting blend values.

The author's studies on blends were done in the mid-1960s when little had been published on the subject. Since then much more attention has been paid to the properties of blends, and reasons have been deduced for the deviations from the values expected based on relat ve proportions of the polymers present. It is now realized that vulcanizate properties are determined in part by the distribution of the filler in the phases of the blends and by the amount of covulcanization— that is, the interphase crosslinking that occurs between the different phases of the blend. For example, carbon black appears to be wetted better and thus to concentrate more in one polymer than another in the blend. This whole area appears to be a fruitful field for further research. A review of elastomer blends is given by Corish [1].

II. POLYBUTADIENE RUBBER

A. Properties

Because of the ease with which natural rubber blends with other elastomers, it has been called the master mixer. A close competitor in that regard would be polybutadiene. By itself polybutadiene is somewhat difficult to mix, so it is usually blended with other rubbers for its desirable properties. It has been used alone—some play balls with exceptional resilience were made with this polymer. BR is the ASTM abbreviation for butadiene rubbers.

Polybutadiene is a homopolymer of butadiene (C_4H_6). It can be made either by solution or in emulsion polymerization; most is made by solution. Butadiene polymerizes by addition, either the vinyl 1,2 form or by 1,4 addition.

Figure 4.1 shows the five ways in which the butadiene unit can join in the polymer chain. In the common ,4 addition there are the cis and trans isomers. In

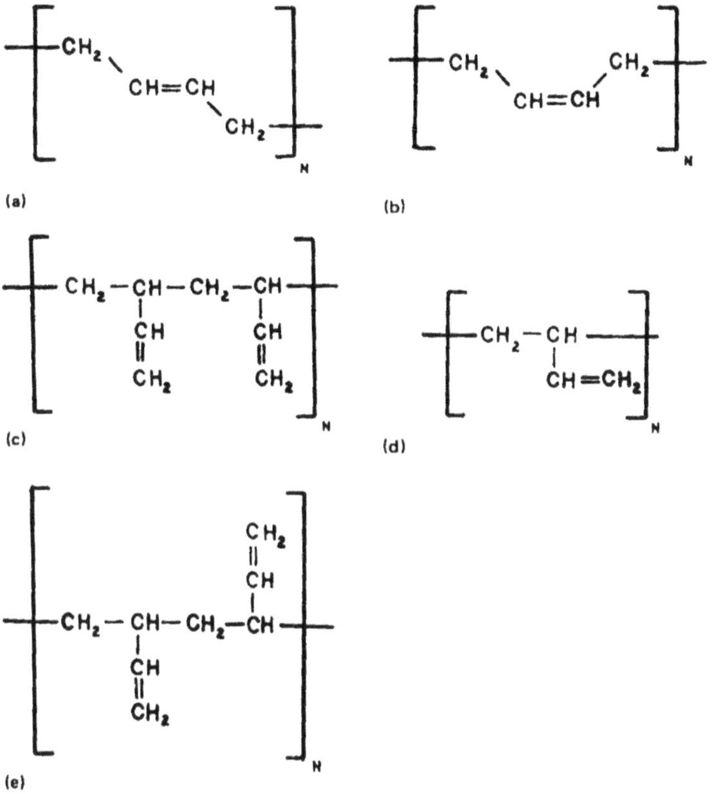

Figure 4.1 Polybutadiene structures: (a) *trans*-1,4 addition, (b) *cis*-1,4 addition, (c) syndiotactic vinyl 1,2 addition, (d) isotactic vinyl 1,2 addition, and (e) heterotactic or atactic vinyl 1,2 addition.

the vinyl 1,2 addition the isotactic structure is that in which all vinyl groups are attached to a backbone carbon in the same spatial arrangement. In the syndiotactic form the vinyl groups alternately attach to the backbone carbons, in the so-called D and L configurations. In the heterotactic or atactic configuration the vinyl groups randomly attach to the backbone carbons.

Depending on the choice of catalyst system solution, polymerized polybutadiene can be prepared ranging from almost 100% cis to 100% trans or 100% vinyl. Since emulsion polybutadiene is rarely used, only the solution-polymerized form is reviewed in this chapter. Most brands of this polymer are called high cis BR; they have cis contents of 95% or more, and a cobalt catalyst usually is used in their production. The low cis varieties normally range from 35 to 38 cis content and are made with a lithium catalyst. There are several oil-extended polybutadienes, plus a limited range containing oil and black.

A difficulty with ordinary low cis BR has been its tendency to exhibit cold flow, the propensity to creep or flow very slowly at ambient temperatures. To get around this hindrance in packaging and storage, a nonflow grade of this type of polybutadiene is available. Its stability is due to a small amount of an incipient crosslinking monomer present during polymerization. This ingredient causes some crosslinking, which reduces the cold flow but does not result in gels or reduce processibility in other ways.

Because of the regularity of its structure, polybutadiene has a tendency to crystallize. This depends on the amount of cis, trans, and vinyl forms present. For example, high cis BR crystallizes at about −40°C and at room temperature when stretched over 200%. On the other hand, rubber with a high trans content of 70–80% is crystalline over a wide range of temperature. Whereas high cis polybutadiene is a soft, easily soluble polymer, high trans rubber is soluble with difficulty in many solvents; it is a tough thermoplastic polymer with high hardness. With a density of 0.91–0.92, polybutadiene is similar to NR and SBR.

There is a relationship between the microstructure of polybutadiene and the glass transition temperature—specifically, the 1,2 addition content of the polymer. At about 10% 1,2 addition, the glass transition temperature is about −95°C. At 95% it is about −15°C It is practically a straight-line relationship. Such information is of interest to tire compounders because the skid resistance of tire treads increases as the glass transition point of the polymers used therein is higher.

Several other relationships are dependent on the cis/trans ratio. An overall review of the changes in physical properties in going from 95% cis to nearly 100% trans has been reported by Crouch and Short [2]: polybutadiene with high cis content is low in uncured tensile strength or green strength compared to natural rubber. This is an important consideration in rubber products that are built up, like tires and hand-made hose, where the parts have to adhere together prior to vulcanization. cis-Polybutadiene usually requires less sulfur than trans-polybutadiene. As the trans content increases, both modulus and heat buildup increase. The test recipe developed by the ASTM to evaluate non-oil-extended polybutadiene is D 31 89-90 [3], reproduced in Table 4.1.

An indication that difficulties in mixing can be expected when the polymer is tested alone rather than in a blend is that four different mixing procedures are offered. Probably an easier formula to test polybutadiene would be one that has been used by Goodyear [4]:

Polybutadiene	90.00
No. 1 RSS	10.00
ISAF black	50.00
Circosol	10.00

Paraffin	2.00
Wingstay 100	1.00
Stearic acid	3.00
Zinc oxide	3.00
Altax	0.45
DPG	0.75
Sulfur	1.25
	171.45

In the recipe above Circosol is a petroleum plasticizer marketed by Sun Oil, Wingstay 100 is a Goodyear antioxidant, and Altax is an accelerator sold by the R. T. Vanderbilt Company; ISAF black is intermediate superabrasion furnace black, and DPG is diphenylguanidine. Reproducibility of this recipe might be improved by using a more closely controlled natural rubber, like SMR CV. Typical vulcanizate properties with a Goodyear polybutadiene, Budene 1207, are given in Table 4.2.

Table 4.1 ASTM Test Recipe for Non-Oil-Extended Polybutadiene

Component	NIS T SRM No.	Quantity (parts by mass)
BR		100.00
Zinc oxide	370	3.00
Sulfur	371	1.50
Stearic acid	372	2.00
Oil furnace black	378	60.00
TBBS[a]	384	0.90
ASTM type 103 petroleum oil		15.00
		182.40

[a]*N-tert*-Butyl-2-benzothiazolesulfenamide.

Table 4.2 Typical Polybutadiene Vulcanizate Properties

Property	Value
300% Modulus, psi (kg/cm^2)	1000 (70)
Tensile strength. psi (kg/cm^2)	2350 (165)
Elongation, %	540
Shore A hardness	59
Tear strength, die C, lbf/in (kN/m)	720 (126)

B. Manufacture

Again only the most popular process, the solution process, is considered here. The microstructure of polybutadiene is largely influenced by the catalyst system used. At least nine such systems have been employed in commercial production. However only two are in primary use today, identified by their metal component as cobalt and lithium catalysts. In the case of cobalt, soluble catalysts are used such as cobalt octonate and naphthenates. Cobalt-catalyzed polymerizations are used to produce high cis polybutadienes—that is, with cis contents in the mid-90s. Lithium-catalyzed polymerizations result in polybutadienes with cis contents averaging perhaps 34%. A commonly used lithium catalyst is butyllithium.

Butadiene was formerly obtained largely from the dehydrogenation of butane and butenes but now is largely recovered from the C_4 stream in petroleum-cracking operations designed to produce ethylene. Very high purity of both monomer and solvent is essential. For example, the butadiene has to be at least 99 mol % pure. Since traces of air and moisture are harmful to the process, the butadiene and solvent used are dried before entering the reactor. Solvents may typically be aromatic, aliphatic, and alicyclic hydrocarbons. Catalysts are charged into the reactor under inert atmospheres.

The polymerization process is a continuous one and is carried on in a reactor fitted with agitators; process temperature control is through either reflux or jacketed cooling. The reaction is carried out to a set time and temperature. Temperature control is necessary because free radical polymerization of butadiene is a function of the reaction temperature. As the polymerization temperature is lowered the cis content decreases. There is no *cis*-1,4 content made below $-15°C$.

At the targeted conversion, a short stop is added to the reactor effluent which stops the polymerization by deactivating the catalyst. Other ingredients such as process oil may be added with the short stop. The catalyst is removed and an antioxidant added to stabilize the polymer. In a subsequent column, steam stripping removes solvent and unreacted monomer, which are recycled. The polymer crumb is dried and baled.

C. Grading

Solution process butadiene rubbers are listed in the *Synthetic Rubber Manual* published by the International Institute of Synthetic Rubber Producers. The 1989 edition of the manual lists no fewer than 26 producers of this kind of rubber with over 100 individual brands listed. Cis content ranged from 98 to 10%, vinyl content from 70 to 0.8%. Mooney viscosity (ML 1 + 4, 100°C) ranged from 32 to 70. Twenty-four brands were oil-extended, usually with 37.5 parts of highly aromatic oil. Three brands were oil-extended black masterbatches with black loadings (ISAF or HAF blacks) going from 77 to 97 parts per 100 parts of rubber.

D. Compounding

Polybutadiene rubbers are usually blended with NR or SBR and find their largest application by far in the tire industry. Here they can impart improved abrasion resistance, resilience, low-temperature flexibility, and increased groove-cracking resistance. Typical applications include passenger tire treads and sidewalls and bias truck tire body plies. About 25–35 wt % BR might be used in passenger tire treads. If the polybutadiene content exceeds 50%, mixing and processing troubles may occur.

Although peroxide curing is possible, polybutadiene as an unsaturated rubber is usually vulcanized by conventional curing systems. BR requires less sulfur for vulcanization than natural rubber, but higher acceleration is necessary to achieve the same state of cure. Going from 100 NR to 100 BR, sulfur dosage might drop from 2.25 to 1.5, but accelerator could increase from 0.5 to 1.0 part. Polybutadiene vulcanizes more slowly than natural rubber and is about equal to SBR.

BR can be compounded with large amounts of conventional fillers and process oils. Usually the fillers are carbon blacks, and in blends the black–polymer phase distribution is of great importance. The effects show both in the rheological properties of the stock as mixed and in its vulcanizate properties. In an interesting paper, Hess and his coworkers [5] worked with a 40-part black loading of ISAF black in a 50/50 blend of NR and BR. They found that with 60% of the black in the BR phase, tensile strength peaked and abrasion resistance continued to improve with increasing carbon black content in the BR. Softeners for polybutadiene rubbers are usually highly aromatic and naphthenic petroleum oils. Although BR is somewhat resistant to oxidation, up to 2 parts of age resisters, mainly antiozonants, usually is added.

Brown [6] gives some of the properties of a BR/SBR tire tread stock for an "all-season" tire. A common emulsion SBR, 1712, which has 37.5 parts of oil with 100 parts of polymer, was used at 103.13 parts with 25 parts of solution BR 1203 (regular IISRP number) to give 100 parts of polymer. This was compounded with 70 parts of an ISAF black and 10 parts of oil together with curatives and age resisters.

Tensile strength was 16.4 MPa (2340 psi), 300% modulus was 8.4 MPa (1200 psi), and ultimate elongation 470%. Shore A hardness, was 63. Tear strength by ASTM method D 624, die B at 50°C, was 60.8 kN/m (540 lb/in.), using die C at the same temperature it was 32.3 kN/m (290 lb/in.). Because tire treads in service operate at higher than atmospheric temperatures, tests at an elevated temperature are more relevant.

III. CHLOROPRENE RUBBER

A. Properties

Neoprene is the popular name for polymers of chloroprene, 2-chloro-1,3-butadiene. Its color depends on the type; it can be light amber, creamy white, or silver gray. Also depending on the type is its density, which is 1.23–1.25, the highest of all synthetic rubbers. Mooney viscosity (ML 1+4 min at 100°C) ranges from about 34 to 130 with lots from an individual type ranging from 5 to 20 units. Neoprene was introduced as Duprene by the DuPont company as a synthetic rubber with good oil, ozone, and weathering resistance. Since it was introduced its resistance to oil has been bettered by nitrile rubbers, its ozone resistance by ethylene propylene terpolymer. It is still used in significant volume because of its good combination o´ properties, its processibility, and the variety of polymers that can be produced from chloroprene.

Neoprene consists mainly (88–92%) of *trans*-1,4-chloro-2-butenylene units as shown in Figure 4.2. The *cis*-1,4 addition accounts for 7–12%, the 1,2 around 1.5%, and the remainder is 3,4 addition. With the 1,2 addition there is some chlorine in the allyclic form. This chlorine is quite reactive and permits neoprene to vulcanize under a variety of conditions without the use of sulfur or sulfur-bearing compounds.

Neoprene owes its resistance to ozone attack, oxidation in air, or weathering to the electronegative chlorine on the 1,4-*trans* double bond. It surpasses other diene rubbers such as SBR or natural but nevertheless requires age resisters to give the best service. Neoprene i considered to be nontoxic. This polymer has flame resistance. and in many cases neoprene compounds ignited with a flame will stop burning when the flame is removed.

An important characteristic of neoprene types is their tendency to crystallize. Because of the regularity of its structure, neoprene crystallizes fairly easily; it has a freezing point of about 14°F. The degree of crystallization in neoprene is largely dependent on the amount of trans configuration in the polymer. Since neoprene prepared between 0 and 40°C is predominantly *trans*-1,4, both the raw and the cured polymer crystallize, especially when stretched. It can be made to

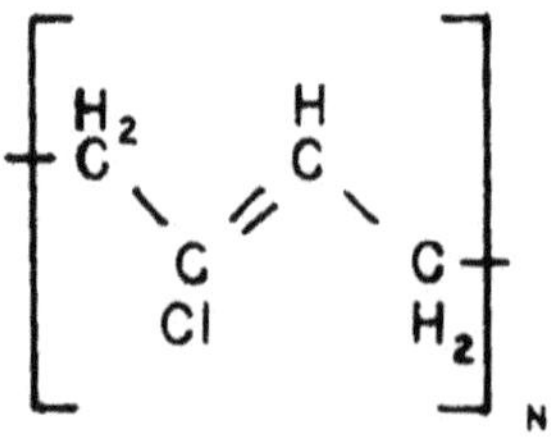

Figure 4.2   2-Chloro-2-butenylene unit (*trans*-1,4 addition).

crystallize more slowly by breaking up the regularity of the chain. This is done by copolymerizing a small amount of dichlorobutadiene. A major division of neoprene polymers is between general-purpose grades used in various elastomeric products and adhesive grades, which are particularly useful in quick setting and high bond strength adhesives.

B.   Manufacture

Neoprene is made by free radical emulsion polymerization. Formerly chloroprene was obtained by the chlorination of acetylene, but for cost and other reasons it is now prepared by chlorinating butadiene. Chloroprene is a liquid at room temperature; its boiling point at atmospheric pressure is 59.4°C.

If chloroprene is polymerized without any modifiers the product is so tough and insoluble that it is impossible to process. The polymer chain must be broken up to allow points for cleavage. There are two methods used to obtain processibility:

1.   The sulfur- or thiuram-modified process
2.   The non-sulfur-modified process

In the first case if a small amount of sulfur is included in the emulsion formula recipe, it copolymerizes with the chloroprene to form a hard plastic material. However, if while still in the latex stage a material like tetraethylthiuramdisulfide is introduced and the blend heated, the polychloroprene becomes peptized and workable. The chemistry of this reaction is not fully understood. Much information on the second process is proprietary. It is known that some manufacturers use mercaptans as regulators of molecular weight.

Neoprene can be made by either a batch process or a continuous process. The latter is of limited advantage if the type of polymer is changed frequently.

In making sulfur-modified rubber the chloroprene is emulsified in water with a rosin acid soap, the emulsion containing about 38% by weight chloroprene. By dissolving wood rosin in the monomer with the sulfur and adding sodium hydroxide and sodium alkyl naphthalene sulfonates, a soap is formed in the emulsion and the sodium alkyl naphthalene sulfonate stabilizes the latex. The reactor charge is emulsified by pumping it through a centrifugal pump several times, and it is then discharged into a glass-lined reactor equipped with glass-coated agitator. Temperatures of 40°C or thereabouts are maintained inside the reactor by having brine at −15°C circulate through a jacket surrounding the reactor. The polymerization is stopped at about 90% conversion by adding an aqueous solution of tetraethylthiuramdisulfide. The reactor contents are cooled to about 20°C and aged, perhaps for 8 hr. In this aging period the tetraethylthiuramdisulfide cleaves the polymer at the sulfur bond, peptizing it. After aging, the reaction mixture is acidified with 10% acetic acid. Acidification serves two purposes: it stops the peptizing action and converts the sodium resinate into rosin acid. The rosin acid then dissolves in the polymer. The latex is now ready for isolation and removal.

The separation, coagulation, and packaging of neoprene is unique. Essentially the latex is continuously coagulated on a revolving frozen drum from which the latex is separated as a film, unfrozen, washed, dried, and twisted into a rope, which is cut into sections and bagged.

The freeze roll, which may be 9 ft in diameter, slowly revolves, partly immersed in a pan of the latex. By circulating brine inside it, the freeze roll can be kept at temperatures of −10 to −15°C. The layer of latex on the roll freezes into a film, which is removed by a stationary knife and is supported on a woven stainless steel belt. Here the web is thawed and washed with water. Squeeze rolls then reduce the water content to perhaps 25% of the dry weight. Hot air at 120°C dries the polymer to negligible amounts of water. The film is cooled to 50°C and sent to a roper, where the polymer is cut into sections and bagged.

At the usual conversion stage say 90%, chloroprene polymers before the tetraethylthiuramdisulfide is added are over 90% gel. The action of that material must be carefully monitored because it could destroy the gel completely. An important factor in the crystallinity of neoprene is the polymerization temperature. The lower the temperature, the greater the tendency to crystallize.

C. Grading

The 1989 edition of the *Synthetic Rubber Manual* mentioned previously named nine suppliers of polychloroprene dry rubber. The IISRP does not assign regular institute numbers to these rubbers but identifies them by supplier's type name and gives data on their Mooney viscosity and a ranking on their crystallization rate. The types available are numerous: DuPont as of 1989 was marketing 27 types. Graff [7] has identified comparable types from various suppliers.

As mentioned earlier, a first subdivision of neoprene types is between those supplied largely for adhesives and the general-purpose types. The latter are by far the most widely used in compounding and are divided into three classes: G, W, and T.

The G types were developed first and were copolymerized with sulfur and thiuramdisulfide stabilizer to make them processible. Typical G polymers are GN, GNA, and GT. Differences between these include the choice of antioxidant (ST or NST) and relative resistance to crystallization. Since neoprene GN has only fair raw polymer stability, inventories should be monitored closely and stocks used shortly after receipt if possible.

The more crystallization-resistant W types are modified during manufacture with a mercaptan, which gives them greater storage stability. They have a more uniform molecular weight distribution. These grades normally require organic acclerators to cure reasonably fast. Typical grades are W, WHV, and WK. WHV is noteworthy because of its high Mooney viscosity (ML 1+4, 100°C) of over

100. It can be extended with large volumes of black and oil to obtain economical compounds. Neoprene WRT is the most crystallization-resistant neoprene.

The T grades are closely related to the W grades and contain a crosslinked polymer gel fraction of polychloroprene to improve behavior. They are crystallization-resistant and do require organic accleration. These types have very low nerve and shrinkage. Typical grades are TW and TRT.

D. Compounding

Neoprene seldom is chosen over other polymers (on the basis) of technical factors alone. Compared to such rubbers as NR and SBR, it is more expensive per pound. In addition, its density of 1.23–1.25 is significantly higher than that of other elastomers, which means higher costs on a volume basis. Accordingly, if oil resistance is wanted, a blend of NR and a butadiene acrylonitrile rubber might be considered. If high ozone resistance is wanted, a blend of NR and EPDM (see Chapter 6, Section III) would be appropriate. This works both ways of course. Addition of BR to neoprene will improve its flex life; similarly, if EPDM is added, ozone resistance is bettered, yet the good processibility of neoprene can be retained. Such blends are not really homogeneous but heterogeneous. It has been theorized that certain technical advantages, such as resistance to cracking, are achieved because the crack terminates at a polymer phase boundary. There is evidence that carbon black is not evenly distributed through a blend but is concentrated more in one of the polymers, which might explain this phenomenon.

Neoprene is used frequently in belts, wire and cable, hose components, and molded or extruded items. It is used in conveyor belts when parts transported or ambient conditions are oily. Its oil resistance is valuable in the tubes and covers of hoses like gasoline curb pump hose and oil suction and discharge hose. Service entrance wire and drop wire require the resistance to abrasion and weather that neoprene provides. Molded pipe gaskets and extruded building seals are made of neoprene.

Neoprene compounds used in outdoor exposures are usually black or dark brown, since light-colored compounds darken in sunlight. On the plus side, neoprene gives strong, pure gum stocks with high tensile strength and adheres well to other materials; its resistance to fire can be a unique advantage. Neoprene vulcanizates commonly meet the requirements of type BC in ASTM D 2000, "Standard Classification System for Rubber Products in Automotive Applications."

After neoprene has been chosen as the polymer, the particular type must be selected. One guideline here can be the Mooney viscosity. It is known that the higher the Mooney viscosity, the better the acceptance of fillers and oils, the better the dimensional stability of uncured shapes, and the higher the tensile strength. Lower Mooney viscosity means lower mixing temperature, lower energy consumption, and better calenderability. This property set is expanded on in Bayer pamphlet E88-773/63005.

The choice of neoprene type usually revolves also around whether a sulfur-modified or non-sulfur-modified type would be most suitable. If such properties as high tensile and tear strengths, good flex-cracking resistance, and high resilience are wanted, the sulfur-modified neoprenes would be chosen. These types masticate more easily but have limited storage stability. The non-sulfur-modified neoprenes offer such advantages as low compression set and high heat resistance; they do require organic accelerators to give practical rates of cure.

Metallic oxides are considered to be essential in curing neoprenes; 4 parts of light-calcined magnesium oxide (magnesia) and 5 parts of zinc oxide is the standard combination. The type of magnesium oxide used is critical, however: it must be precipitated, not ground and calcined after precipitation, to ensure high activity. It also should be protected from moisture absorption. The foregoing combination of oxides usually suffices for the G grades, although thiourea can be used to speed up the cure if desired. The W grades do require organic accelera-tion as well, and thiourea at 0.5–1.0 part is probably the most economical.

Antioxidants are necessary in all neoprene compounds to ensure good aging. Formerly phenyl-β-naphthylamines were used, but because of possible toxicity, para-oriented styrenated diphenylamines like Goodyear's Winstay 29 are recom-mended. Dosage is usually 1–2 parts on the rubber. If antioxidant protection is needed, diaryl *p*-phenylenediamines should be used.

The G and W neoprene types, which are used for most applications, do give different properties with the same fillers and reinforcing agents. At the same loading, vulcanizates of neoprene W not only are softer than their GN counter-parts but they generally have higher tensile strengths. Ordinarily, neoprene com-pounds use thermal blacks, clays, and fine particle whitings. If high tensile strength is demanded, semi-reinforcing furnace (SRF) blacks can be used, blacks with finer particles (e.g., HAF) providing only marginally more strength and frequently giving stiff, scorchy stocks. Whiting is an extender in neoprene; soft clays show a minor reinforcing action, hard clays still more. Precipitated silica or a hydrated calcium silicate is used for light-colored stocks and with high tensile and tear strengths. Abrasion resistance is higher, but dynamic properties are poorer.

The usual softeners for neoprene are low cost petroleum oils, commonly aromatic or naphthenic. Paraffinic oils are resisted by neoprene, and they exude. Most stocks require 10–20 parts of softener per 100 parts of neoprene for processing. If the filler loading is such that this range of softener is not exceeded, a naphthenic softener can be used which neither darkens light-colored vulcani-zates nor stains adjacent surfaces.

Processing aids such as low molecular weight polyethylene, microcrystalline waxes, and palm oil improve release from molds, help extrusion, and aid in dispersing fillers and reinforcers. To tackify neoprene compounds, coumarone indene resins and hydrogenated rosin esters are frequent choices. Good low temperature properties are achieved by using ester-type plasticizers like dioctyl sebacate. For retaining the flame-resistant properties in neoprene compounds, nonflammable phosphoric acid esters or chlorinated hydrocarbons can be used as replacements for mineral oils.

A basic problem in mixing neoprene stocks, especially the sulfur-modified ones, is the prevention of scorch. With neoprene, the effect of heat on the uncured stock is cumulative. Mixing temperatures should be kept low ($<$ 120°C). The magnesia, which retards the scorching tendency, should be added early in the mixing cycle—however, not before a mill band has formed or the neoprene has massed in an internal mixer, since the magnesia may cake on cold metal surfaces. Zinc oxide and accelerators are added last in the mixing cycle. If this would cause scorching, they should be added on the warm-up mill or held until the stock is to be processed further. Zinc oxide or accelerators often can be added masterbatched in SBR.

REFERENCES

1. Corish, P. J., Elastomer Blends, *Science and Technology of Rubber*, F. R. Eirich, ed., Academic Press, New York, 1978, p. 489.
2. Crouch, W. W., and Short, S. N., *Rubber Plast. Age*, *42*, 276 (1961).
3. D 3189-90, Rubber—Evaluation of Solution BR (Polybutadiene Rubber), *1991 Annual Book of ASTM Standards*, Vol. 09.01, p. 519.
4. Kuzma, L. J., in *Rubber Technology*, 3rd ed., M. Morton, ed., Van Nostrand Reinhold, New York, 1987, p. 253.
5. Hess, W. M., Scott, C. E., and Callan, J. E., *Rubber Chem. Technol.*, *40*, 371 (1967).
6. Brown, T., *The Vanderbilt Rubber Handbook*, 13th ed., R. T. Vanderbilt, Norwalk, CT, 1990, p. 89.
7. Graff, R. S., in *Rubber Technology*, 3rd ed., M. Morton, ed., Van Nostrand Reinhold, New York, 1987, p. 339.

5

Elastomers: Butyl and Ethylene Propylene Rubbers

I. SCOPE

This chapter covers butyl rubber, a special-purpose rubber that nevertheless is rather widely used, and ethylene propylene rubbers, the latest type of synthetic rubber with significant volume. Butyl's outstanding property is its low permeability to gases. Butyl rubber can be halogenated, and chlorobutyl and bromobutyl are available. These rubbers are useful in such applications as tire inner liners, white sidewalls, and pharmaceutical closures. Ethylene propylene rubbers are of two types: copolymers and terpolymers. The copolymers, consisting only of ethylene and propylene, are saturated and therefore cannot be vulcanized by conventional means. Since the terpolymers also contain a small amount of a diene, which provides some unsaturation, they can be vulcanized. The terpolymer is by far the most useful and will be given the most consideration. An outstanding property of terpolymers is the high loading of oil and fillers that can be used with them, and yet the compound shows rubbery properties. Because of their relative cheapness, ability to be loaded, and aging resistance, the terpolymers are used in a very wide range of compounds by themselves or in blends.

II. BUTYL (IIR) RUBBER

A. Properties

Butyl rubber is a copolymer of isobutylene with small amounts of isoprene, hence the acronym IIR. Before the introduction of the tubeless tire about 1954, passenger tire inner tubes were made almost entirely from butyl. It had displaced natural rubber in that product because of its very low gas permeability, reducing significantly the frequency of testing tires for adequate inflation.

The low permeability of butyl stocks to air transmission was reported by Dunkel et al. [1]. Driving tests for air loss of inner tubes were conducted on natural rubber and butyl inner tubes. Both sets of tubes were originally inflated to 28 psi. The tests were controlled road tests on cars driven 60 mph for 100 miles/ day. At the end of one week, air pressure loss for the natural rubber tubes was 4.0 psi, for the butyl tubes 0.5 psi. At the end of one month the natural rubber tubes were down to 16.5 psi, butyl down only 2.0 psi. Butyl is still used in the small market for inner tubes, but butyl has found extensive further uses in such items as hose, mechanical rubber goods, and wire and cable insulation. Besides its low gas permeability butyl has excellent resistance to heat, chemical attack, and ozone.

Figure 5.1 shows the structure of butyl rubber with an isoprene unit included. The copolymer contains about 97–98% by weight polyisobutylene and 2–3% by weight isoprene. In its manufacture isoprene can add on to the growing chain in three ways: 1,2; 3.4; or 1,4 addition. Various methods have established that it is the 1,4 addition that takes place Flory found that the molecular weight M of butyl can be calculated from the intrinsic viscosity η in diisobutylene at 20°C by the following equation [2]:

$$\log M = 5.378 + 1.56 \log \eta$$

Butyl has unique properties that are due to its very low unsaturation. Commercial grades of butyl vary from 0.6 to 2.5 mol % unsaturation. Since the molecular weight of an isoprene molecule is 68, the molecular weight between points of unsaturation is 68. On the other hand, a butyl rubber with 1 mol %

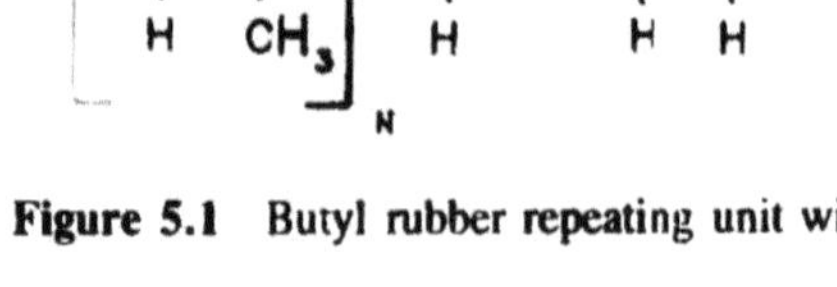

Figure 5.1 Butyl rubber repeating unit with isoprene unit included.

unsaturation would have a molecular weight between points of unsaturation exceeding 5000. It is perhaps somewhat oversimplified but useful to note that if the low unsaturation makes butyl more difficult to vulcanize, it also provides few points of attack for various breakdowns.

A prime example of butyl's resistance to breakdown is in mixing behavior, where it does not break down as other rubbers do on mastication. (However heating the butyl before mixing reduces mixing time.) Diffusion of gases through a polymeric membrane is dependent on the solubility of the gas in the film. Polyisobutylene, by far the major component of butyl, is relatively impermeable to gases compared to other polymers. For example, if permeability to air at 65°C is taken as 100 for natural rubber, it is about 80 for SBR and 10 for butyl. Such relationships should be used with care. Permeability varies with temperature and choice of fillers, among other factors.

Butyl has excellent resistance to ozone and weathering, again because of its low unsaturation. For example, if the lowest mole percent unsaturated types are used, satisfactory weather resistance for single-ply roof sheeting membranes can be obtained by simply adding 3–5 parts of paraffin wax to the mix and no other age resister.

Butyl has a high degree of resistance to water and mineral acids. Although butyl rubbers are swollen by hydrocarbon solvents and oils, they are resistant to polar liquids, vegetable oils, and synthetic hydraulic fluids.

Before leaving this section, a butyl neoprene blend that has been very useful should be noted. Tire curing bladders operate at 350°F or higher, and an early problem was finding a rubber that would stand up to this steam temperature cycle after cycle. Although conventionally vulcanized butyl will not perform adequately under these conditions, if as little as 5 parts of neoprene to 100 of butyl is used in conjunction with a resin-type cure, very satisfactory results are obtained.

Finally, butyl has good low temperature flexibility, and the linkages make up the main chains have high rotational flexibility. Butyl rubbers are gel-free and completely soluble in solvents like benzene or hexane.

B. Manufacture

The manufacture of butyl rubber is continuous, a solution polymerization being used with methyl chloride as the carrier solvent. The monomers isoprene and isobutylene are obtained from streams coming off petroleum-cracking units. Feedstock treatment ensures that isobutylene has a purity of over 90% is more than 90% pure as it enters the process, isoprene 95%. A Friedel–Craft catalyst like $AlCl_3$ (soluble in methyl chloride) is used, often with a cocatalyst. Butyl rubber polymerization differs from other synthetic rubber polymerizations in having the coldest reaction temperature (-90 to $-100°C$). Even at such low temperatures, the reaction is very fast.

To keep the temperature at the desired level, the isobutylene, isoprene, AlCl₃ dissolved in methyl chloride, and the methyl chloride carrier are cooled to near the reaction temperature before being pumped into the reactor. Good catalyst dispersion is obtained by having vigorous agitation in the reactors. The reaction is quite exothermic. Cooling, necessary because of the amount of heat generated, is obtained by boiling ethylene in a jacket around the reactor. Other reaction conditions being equal, the molecular weight of the polymer increases with decreasing polymerization temperature. Depending on the grade of butyl being produced, conversion of up to 95% of the isobutylene (the more reactive monomer) occurs. Residence time in the reactor will vary from 30 to 60 min.

In the reactor the polymer emerges as discrete crumbs. For its recovery the crumb slurry flows continuously out of the top of the reactor into a tank containing hot water. Much of the unreacted monomer and the methyl chloride is flashed off to be recycled. At this point an antioxidant may be added to the rubber. Sodium hydroxide and zinc stearate are also added here. The sodium hydroxide decomposes the AlCl₃ initiator; the zinc stearate prevents the wet crumbs from sticking together. The water slurry flows to a tank, where under high vacuum the remaining monomer and solvent are removed. This is an important step because methyl chloride is a toxic chemical. The raw polymer is now ready for the finishing operations. A major portion of the water is removed by having the slurry flow over vibrating shaker screens. The rubber crumb can then be dried by passing it through dewatering extruders. In this method the elastomer moves through the last extruder at high pressure and temperature, and the remaining moisture flashes off as it leaves the die plate. Alternatively, the rubber can be dried by passing t through roll mills. The final dried product is cooled to baling temperature, weighed, pressed into bale form, and wrapped in polyethylene film, 75 lb to the bale. For shipment it is packed in cardboard cartons or wooden crates.

The versatility of butyl rubbers was increased by the development of halogenated butyls, chlorobutyl and bromobutyl. When bromine or chlorine is joined to the butyl molecule in a 1:1 molar ratio of chloride or bromide to double bond, covulcanization with general-purpose highly unsaturated polymers is possible. A most important application of such rubbers is in the inner liners of tubeless tires. In this application these polymers show very low air permeability and good adhesion to the general-purpose rubbers used in the body of the tire.

Halogenated butyls are made by reacting a heptane solution of butyl with chlorine or bromine at 40–60°C. The hydrogen chloride or hydrogen bromide evolved is neutralized with an aqueous caustic solution. After this solution is removed from the two-phase system, the halogenated product is finished like normal butyl, but a stabilizer is added as well as an antioxidant. The stabilizer prevents hydrogen chloride or hydrogen bromide from continuing to evolve, yet it must not impair the vulcanization ability of the rubber or introduce a health hazard. Calcium stearate and an epoxidized vegetable oil are suitable for this purpose for bromobutyl; calcium stearate alone suffices for chlorobutyl.

Although about 3% of halogen can be reacted with the rubber, in practice, chlorobutyl contains about 1.2–1.3% bound chlorine, brominated butyl about 2.2% by weight bound bromine.

C. Grading

Commercial butyl grades are not as numerous as SBR or neoprene grades. They are listed in the IISRP manual but, like the neoprenes, have no assigned institute numbers. The always important property of whether the stabilizer is staining or nonstaining is given, along with the unsaturation, which varies from 0.6 to 2.5 mol %. Most of the Mooney viscosities are given under the usual test conditions for butyl, 1 + 8 min at 100°C, rather than the 1 + 4 min commonly used for other polymers. The nominal minima are 18–61, the maxima 26–70, each depending on grade. Some butyl grades are approved for direct contact with food as they comply with regulations of the U.S. Food and Drug Administration (FDA) for such products.

D. Compounding

When regular butyls are used for compounding, the appropriate grade is chosen on the basis of the amount of unsaturation in the polymer, then on the Mooney viscosity. For top resistance to ozone, the weather, and flexing, butyls up to 1.0 mol % unsaturation are selected. For general use, 1.5–2.0 mol % unsaturation will give good ozone resistance; for good heat resistance, unsaturation above 2.0 mol % should be satisfactory. As unsaturation increases, ozone resistance decreases, but stocks are faster curing. If low filler and plasticizer levels are used, grades with lower Mooney viscosity should give smoother sheets and extrusions.

With the appropriate curing system, butyls can be vulcanized at temperatures as low as 212°F and as high as 375°F. However these synthetics are usually cured at 300–350°F, somewhat above the customary range for natural rubber. To compensate for the few unsaturated sites available, the more active accelerators (e.g., thiurams and thiocarbamates) are frequently used.

Of the polymers reviewed in this book, butyl utilizes the widest range of curing systems. At least four systems are employed: sulfur, sulfur donor, resin, and quinoid. A general-purpose sulfur-curing system might be 2.0 parts sulfur, 1.5 parts tetramethylthiuramdisulfide, and 1.0 part mercaptobenzothiazole disulfide. For good compression set resistance, and heat resistance, a sulfur donor system could be used (e.g., 2.0 parts dimorpholine disulfide, 2.0 parts tetramethylthiuramdisulfide).

In commercial compounding the quinoid and resin cures are unique to butyl. Quinoid cure systems are useful where excellent age and heat resistance are wanted. A typical system might be 1.5 parts p-quinone dioxime, 4.0 parts mercaptobenzothiazole disulfide, and 5.0 parts Pb_3O_4. There are tradeoffs for the excellent aging this type of cure system gives: despite the very fast cure rate, the vulcanizates are less resilient, and the system is limited to dark-colored compounds. In this method the dioxime is oxidized to dinitrosobenzene. The proposed mechanism is given by Fusco and Hous [3].

Still another vulcanization combination for butyl that gives excellent aging stocks is called the resin cure. Here the crosslinking agent is a derivative such as a methylol phenol resin as shown in Figure 5.2. Vulcanizing is very slow with resin alone, but with stannous chloride dihydrate, the reaction rate is much faster. Proportions for this combination might be 12 parts resin and 3 parts stannous chloride dihydrate. This system is particularly distinguished by the good aging it gives at high temperatures. For this reason it is used for tire bladders.

Vulcanization of halogenated butyl rubber is much easier than with butyl itself. The reaction rate is much faster, and crosslinking can go with either the allylic hydrogen or the halogen or with both. Because of the wide variety of curative systems used, not only with butyl but with the halogenated types, it is impractical to discuss them in detail here. Detailed information can be found in the review by Fusco and Hous [4].

The highest tensile strength is obtained with pure gum butyl stock rather than carbon black reinforced mixes. However, carbon blacks and inorganic fillers do reinforce in the sense that modulus is increased as well as tear strength. Carbon blacks provide the most reinforcement; next in line are such inorganic pigments with finer particles as clay, whiting, and finely divided silica. Some trouble may be experienced in getting suitable dispersions of these inorganic fillers in the butyl matrix. Curative dosage should be adjusted for the adsorptive effect of fine particle silica if used; the amount is found by experiment. Hard clay should be avoided if high processing temperatures are involved as it can cause depolymerization.

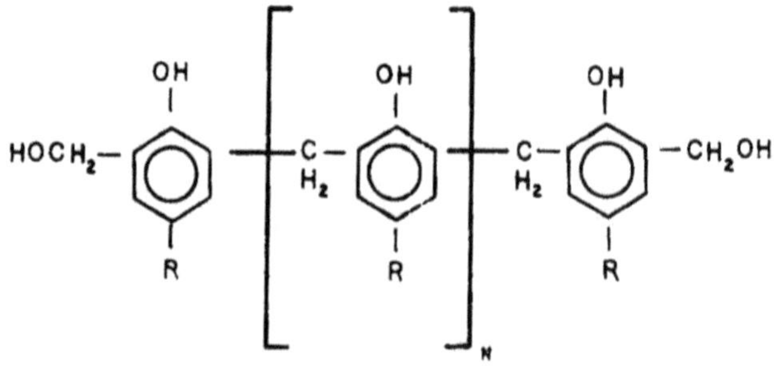

Figure 5.2   2,6-Dimethylol-4-alkyl phenol resin, R-alkyl group.

For softening and plasticizing, butyl stocks require highly saturated, low polarity materials such as paraffinic oils, waxes, aromatic esters, or hydrogenated ester gum. Unsaturated materials would compete for the vulcanizing agent. With normal, low viscosity butyls, 2–5 parts of softener might be used. Higher viscosity rubbers or compounds for purposes like frictioning require significantly higher amounts. Metallic soaps such as zinc stearate help produce smooth compounds, diminishing nerve and tack. Zinc oxide is necessary for good vulcanization, and 5 parts is a common dosage.

Butyl stocks, with their low unsaturation, cannot be satisfactorily blended with such highly unsaturated rubbers as SBR or NR. Precautions should be made in the factory storage of butyl stocks to ensure that the possibility of mixing with other stocks is remote. Salvage is difficult or impossible. One development in compounding not duplicated with other polymers is the heat treatment or thermal interaction of reinforced or loaded stocks. This treatment imparts high elasticity and resilience to the vulcanizate [5].

III. ETHYLENE PROPYLENE RUBBERS (EPM AND EPDM)

A. Properties

Ethylene propylene polymers are the fastest growing synthetics in commercial production. Two broad varieties are made: copolymers and terpolymers. The copolymers are commonly referred to as EPM rubbers, where E and P naturally stand for ethylene and propylene, and M indicates that the rubber has a saturated chain of the polymethylene type. The terpolymers are similarly referred to as EPDM rubbers, where E, P, and M have the same meanings as above and D means that units derived from a diene rubber are in the polymer, with the residual unsaturation in a side chain. The copolymers composed of ethylene and propylene alone are high molecular weight, amorphous, saturated polymers. To produce more useful crosslinked polymer chains, these saturated compounds must be treated with agents like organic peroxides or chlorine, or with radiation, to introduce chemical links between the chains. Terpolymers, vulcanized by conventional means, make up 95% of the market, copolymers the remaining 5%. A diolefin has been copolymerized with the ethylene and propylene, providing unsaturation for crosslinking in side chains pendant from the saturated backbone.

Ethylene propylene-based polymers have received such wide acceptance that they were the subject of an international conference at the University of Leuven, Belgium, in April 1991. A review of the papers has been given by Bowtell [6]. Some points from the 27 papers presented at the conference follow.

Worldwide consumption of ethylene propylene rubber in 1990 was estimated at 600,000 metric tons. Since the Western European market was estimated at 205,000 metric tons in 1990 and predicted to rise to 250,000 in 1995, the rate of increase is quite high. Two newer methods for crosslinking EPDM are using coagents with peroxides and electron beam curing. Several papers at the conference dealt with processing differences in EPDM caused by differing molecular weight distribution (MWD). In extrusion applications, polymers with a broad molecular weight distribution gave better surface appearance at high extrusion rates, with high levels of scorch safety. However narrower MWD polymers have better cure performance. To solve this dilemma, Exxon Chemical has developed a polymer with narrow MWD but with set amounts of high and low end molecular weight polymer to increase processing latitude. A measure of the molecular weight distribution is in the ratio of the weight-average molecular weight to the number-average molecular weight $\overline{M}_w/\overline{M}_n$).

Polymerization is accomplished using Ziegler–Natta catalysts, which induce polymerization by an ionic mechanism. 1,4-Hexadiene (1,4-HD), dicyclopentadiene (DCPD), and ethylidene norbornene (ENB) have been used as the third polymer; the most widely used is ENB. Oil-extended varieties of EPDM are available with up to 100 parts of paraffinic or naphthenic oil. Ethylene propylene rubbers have a Mooney range from 20 to 80 ML (1 + 4) at 125°C. EPM and EPDM rubbers are colorless and ordinarily have a density of 0.86, although this may increase to 0.87 or 0.88 depending on the density and amount of oil added.

The structures for ethylene propylene copolymer and the terpolymer are shown in Figure 5.3.

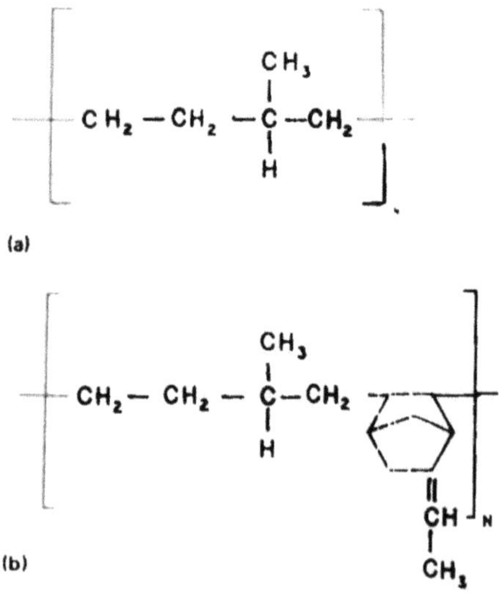

(a)

(b)

Figure 5.3 Ethylene propylene copolymer and terpolymer. (a) Copolymer unit; ethylene and propylene units may be in other than the 50:50 ratio shown. (b) Terpolymer unit, with the same qualification as (a); ethylidene norbornene (ENB) added as the diene to provide unsaturation.

Both polyethylene and polypropylene are rigid plastics at normal temperatures as a result of crystallization. Copolymers of the two do not crystallize, however, because the linear regularity of the homopolymers is not matched and each type of unit retards the crystallizing effect of the other. The addition of the third monomer in EPDM does not change the copolymer properties very much in processing.

Commercial EPDMs contain 45–75% by weight ethylene. Because of crystallization effects, ethylene propylene rubbers high in either ethylene or propylene have higher tensile strength and elongation than rubbers with approximately equal amounts of each monomer. As the ethylene content increases in these rubbers, the rebound resilience increases. Copolymers having an ethylene/propylene weight ratio much above 70:30 are not very rubbery. Besides higher resilience, ethylene propylene rubbers with higher ethylene content give vulcanizates of higher modulus, hardness, and tensile strength.

EPM and EPDM rubbers have a brittle point of $-95°C$ and a glass transition temperature of $-60°C$. These rubbers, particularly EPDM, have numerous advantages which explain their wide and growing use. Their densities are low, they process easily on rubber machinery, and they blend well with other polymers. Their heat resistance is good and they are flexible at low temperatures. They have excellent resistance to ozone, sunlight, and many chemicals. More than any other polymer, they can be extended with oils, fillers, and reinforcing agents yet retain their rubbery properties. On the other hand, they do not give good pure gum vulcanizates and their resistance to aromatic hydrocarbons is poor. The largest use is probably in nontire automotive products. Recently, use of rubber membranes for roofing have been growing.

B. Manufacture

Manufacturing processes for the EP rubbers are proprietary and the subject of many patents. Accordingly, process details such as exact times and temperatures, catalysts used, and monomer purification are not freely available. What follows might be considered to be a general outline of the process.

Both ethylene and propylene are made in huge volume in the United States. For many popular grades the composition is 65–75 mol % ethylene. Many monomers have been tied as the third ingredient in the terpolymer but the three mentioned previously—1,4-hexadiene, dicyclopentadiene, and ethylidene norbornene—are now used exclusively. The first is proprietary to DuPont, which produce ethylene propylene rubbers under the trade name Nordel. Copolymer Rubber and Chemical Corporation first commercialized the use of ethylidene norbornene in their EPsyn trademarked rubbers in 1967, and this monomer is used by many other manufacturers.

Solution polymerization is a common method for manufacturing EP polymers. A prime requirement is the dryness of the monomers and the carrier solvent; otherwise the catalyst could be inactivated by a polar material like water.

To achieve dryness, molecular sieves are used for the monomers; the carrier solvent (often hexane) can be redistilled to remove impurities. The process is a continuous one, and catalysts such as transition metal halides and metal alkyls are used. Both monomers and solvent are refrigerated before entering the reactors, and all components are very carefully metered. Heat from the exothermic polymerization reaction is removed so that the temperature is held at about 35°C. Polymerization is very fast.

By 10% conversion of the monomers, the solution viscosity is so high that further increases would lead to problems, and the polymerization is stopped by having the rubber slurry continuously overflow from the reactor to a flash tank at reduced pressure. This removes most of the unreacted monomers and the hexane, which are recovered for recycling. Catalysts are separated from the rubber slurry.

The rubber, as floc, is now suspended in hot water at perhaps 3–4% concentration. The floc is thoroughly washed, and the bulk of the water removed, by having the slurry flow over shaker screens. The remaining water is removed by having the floc pass through mechanical screw dewaterers. These may be in sequence, with the second one heating up the rubber so much that as it emerges from the perforated die plate, the remaining water flashes off. The rubber pellets, now dry, are cooled on a conveyor belt. The product is then weighed, pressed into bale form, wrapped in film and packed in large cartons.

C. Grading

There is a multiplicity of grade types of ethylene propylene rubbers. Once again the IISRP does not assign institute numbers but does list various producers' offerings under the heading "Types." Characteristics listed include the Mooney viscosity and the polymer's staining behavior. Other identifying features given are whether oil has been added to the polymer, the type (paraffinic or naphthenic) and amount, and whether the third monomer is ENB, HD, or DCPD. Finally, the polymer is described by its structure (random, block, or sequenced—relating to the position of the monomer units in the chain); whether the ethylene content is low, medium, or high; and, finally, whether the unsaturation is low, medium, high, or very high.

Typical of the supplier information provided on EPDMs is the following, given by the Copolymer Rubber and Chemical Corporation for their fast-curing, general-purpose terpolymer EPsyn 40A:

Raw Mooney viscosity (ML 1–4 @ 125°C)	40
Specific gravity	0.86
Volatile matter, % max	0.75

Table 5.1 gives additional typical supplier data.

Table 5.1 Some Properties of EPsyn 40A: ASTM D 3568-81, No. 1 MIM Mixed[a]

Quality control formulation[b]			
Component	Parts	Typical rheometer test values[c]	
EPsyn 40A	100	M_L, dN·m (lbf·in.)	6.8 (6.0)
IRB #5 (N 330 Black)	80.0	M_H, dN·m (lbf·in.)	45.2 (40.0)
Circosol 4240	50.0		
ZnO, NBS 370	5.0	t_s', min	2.0
Sulfur, NBS 371	1.5		
Stearic acid, NBS 372	1.0	$t'90$, min	14.0
TMTD, NBS 374	1.0		
MBT, NBS 383	0.5		
	239.0		

[a]MIM stands for miniature internal mixer; since mill handling of EPDM rubbers is more difficult than is the case with most other rubbers, internal mixing is preferred.

[b]IRB #5 is an industry reference black, the fifth lot so designated, having known properties in rubber; it is an HAF black. Circosol 4240 is a rubber process oil marketed by the Sun Oil Company. The national Bureau of Standards (NBS) provided standard compounding ingredients often used for referee purposes in 1981 when this ASTM standard was written. TMTD and MBT are accelerators; tetramethylthiuramdisulfide and mercaptobenzothiazole, respectively.

[c]Obtained by using an oscillating disk curemeter (e.g. the Monsanto rheometer). M_L is the minimum torque obtained with the compound, M_H the maximum torque, t_s' the minutes required for the torque to rise 1 dN·m above M_1, and $t'90$ the minutes to 90% of the torque increase.

From the rheometer test values for M_L and M_H, the experienced compounder can get a good idea of the processibility of the rubber. The t_s1 and $t'90$ minute values give a good idea of the rate of cure. In production control of EPDM rubbers 160°C is used as the vulcanizing temperature, considerably above 140°C used to test natural rubber and 145°C used for SBR.

D. Compounding

The 1989 edition of the IISRP *Synthetic Rubber Manual* lists more than 150 different types of ethylene propylene rubber. Trouble can be expected if one type is substituted for another, so compounders should review their suppliers' literature and recommendations before deciding what type to use. Some generalizations may be useful. For molding, medium to high propylene content polymer is preferred by most compounders. To have good green strength for shape retention in extruding, high ethylene types are most suitable. For calendering, select a medium to high polypropylene range with a broad molecular weight distribution. Consideration also should be given to using two types of EPDM polymer to achieve the best balance of properties.

As noted earlier, two important and valuable aspects of compounding with EPDM are its ability to be extended with fillers and oil and its use as a polymeric nonstaining antiozonant. The amoun. of extension possible while retaining rubbery properties is greater than for any other polymer. One formula appearing in a supplier's literature called for 250 parts of mixed blacks, 200 parts of whiting, and 200 parts of process oil per 100 parts of rubber. This compound is only about 11.5% by weight polymer yet had a tensile strength exceeding 700 psi. It might be used as a garden hose tube, for example. To impart good ozone resistance when used with other polymers, about 30% by weight of the polymer content should be EPDM. Naturally careful attention must be paid to the curative system to make sure that a good state of cure is obtained.

EPDM batches are often somewhat difficult to mix and usually are mixed in an internal mixer. If mill mixing is necessary, the low viscosity, high propylene EPDMs should be used. Frequently "upside-down" mixing is favored. That is, the rubber is put in after the fillers, oils, and other compounds have been loaded.

Ethylene propylene rubbers use naphthenic oils when used for most applications. However, if the article is used at high temperatures, less volatile oils like paraffinics should be substituted. One problem here is that at low temperatures, say 45°F or lower, paraffinic oils may bleed out. In such cases a blend of naphthenic and paraffinic oils may be satisfactory.

Pure gum EPDM compounds are of little strength and value, so blacks are used for reinforcing. For less reinforcement requirements, nonblack fillers include soft clays, fine particle silicas, and calcium carbonate. The more reinforcing blacks are hard to disperse; if possible medium reinforcing blacks or those with higher structure should be chosen.

Curing systems for EPDM are similar to those used for general-purpose rubbers. There is a wide variation in curing for EPDMs depending on how much ENB (or other diolefin) is present—the more the faster. A general-purpose cure system might be 1.5 parts sulfur, 1.5 parts tetramethylthiuramdisulfide, and 0.5 part 2-mercaptobenzothiazole, all in the usual units (parts per hundred of polymer). If a faster cure rate is wanted, boosting the tetramethylthiuramdisulfide to 2.5 parts might be tried. Filler and oil loadings up to 100 parts of each are not uncommon. Carbon blacks, fillers, or pigments are always included in EPDM compounds.

Besides conventional curing systems, organic peroxides can be used. Of the three monomers commonly chosen ethylidene norbornene gives the fastest cure, followed by dicyclopentadiene and 1,4-hexadiene, as measured by rheometer torque. This subject has been reviewed by Crespi et al. [7].

REFERENCES

1. Dunkel, W. L., Neu, R. F., and Zapp, R. L., in *Introduction to Rubber Technology*, 1st ed., M. Morton, ed., Van Nostrand Reinhold, New York, 1959, p. 317.
2. Flory, P. J., *Ind. Eng. Chem* .38, 417 (1946).

3. Fusco, J. V., and Hous, P., in *Rubber Technology*, 3rd ed., M. Morton, ed., Van Nostrand Reinhold, New York, 1987, p. 325.
4. Fusco, J. V., and Hous, P., in *Rubber Technology*, 3rd ed., M. Morton, ed., Van Nostrand Reinhold, New York, 1987, Chapter 10, p. 284.
5. Dunkel, W. L., Neu, R. F., and Zapp, R. L., in *Introduction to Rubber Technology*, 1st ed., M. Morton, ed., Van Nostrand Reinhold, New York, 1959, p. 325.
6. Bowtell, M., *Elastomerics*, July 1991, p. 6.
7. Crespi, G., Valvassori, A., and Flisi, U., in *The Stereo Rubbers*, W. H. Saltman, ed., Wiley, New York, 1977, p. 365.

Elastomers: Nitrile, Polysulfide, and Recycled Rubbers

I. SCOPE

This concluding chapter on elastomers offers a comparison chart (Table 6.2, below), which should help in selecting the proper elastomer or blend for most applications. Three very different rubbers are considered here, all filling specialized niches. Nitrile rubber was originally called Buna N rubber and later butadiene-acrylonitrile rubber; the term *nitrile rubber* is most commonly used, however, and will be adopted here. The ASTM designation for this rubber is NBR. Polysulfide rubbers are unique in having a one-letter ASTM designation (T). They are manufactured by the Chemical Division of the Thiokol Corporation. Recycled rubber has no ASTM designation. This category consists of reclaimed rubber, which is cured rubber that has been involved in a devulcanizing process, and vulcanized scrap rubber that has been shredded or is hard rubber dust. A common form of reclaimed rubber is tire reclaim (usually a blend of SBR and BR). Reclaimed natural, reclaimed butyl, and reclaimed neoprene are also marketed.

Nitrile rubber is chosen where oil and fuel resistance superior to that of neoprene is wanted or for its chemical resistance. Polysulfide rubber is used for its excellent oil and fuel resistance—say, where even NBR cannot function well, such as for oil suction tubes and discharge hoses that are used to convey benzene. Reclaim finds itself in competition with high black, high oil SBR masterbatches and is used as the sole elastomer in inexpensive noncritical products. It has been used to make cheap but serviceable compounds for pedal rubbers on children's wheeled toys, car mats, and even garden hoses. In blends it gives definite processing advantages such as short breakdown and mixing times and fast extrusion, and it firms up uncured stock. Ground or shredded rubber is added to rubber compounds frequently as an extender and is often used in rubber matting. Hard rubber dust is employed in very hard rubber products like pocket combs, caster wheels, and battery cases.

II. NITRILE RUBBER

A. Properties

Nitrile rubber is a copolymer of butadiene and acrylonitrile, made in an emulsion process like styrene-butadiene rubber, as hot and cold polymer. Nitrile has good resistance to a wide variety of nonpolar oils, fats, and solvents. This property offers advantages in making products for the automotive and oil industries such as oil seals, pipe protectors, and blowout preventers. The acrylonitrile content varies, usually from 20 to 50% by weight, with levels under 25% classified as low, 25–35% as medium, and 35–50% as high. Major properties are dependent on the acrylonitrile (ACN) content, and this subject will be picked up again. Mooney viscosity (ML $1 + 4$ at 100°C) ranges from 25 to over 100, although the usual range is 40–80. Unlike natural rubber, NBR shows no crystallinity and therefore there is no self-reinforcing effect. Density varies with ACN content: low ACN level polymer is 0.98, high 1.00. Dunn and Blackshaw [1] estimate the temperature limit in service as 135°C in the presence of oxygen.

The basic reaction in nitrile production is shown in Figure 6.1. Three types of addition are possible: 1,2; 1,4; and 3,4; the latter two are identical. The major part of NBR is 1,4 addition. Hofmann, studying a 28% bound acrylonitrile polymer, found 89.5% to be 1,4 units, 10.5% 1,2 units [2]. Apart from the type of addition present, one must consider the ratio of *cis*- to *trans*-1,4-butadiene. Dunn and Blackshaw [3] found the following values for a hot NBR (30°C polymerization temperature) and a cold NBR (7°C polymerization temperature).

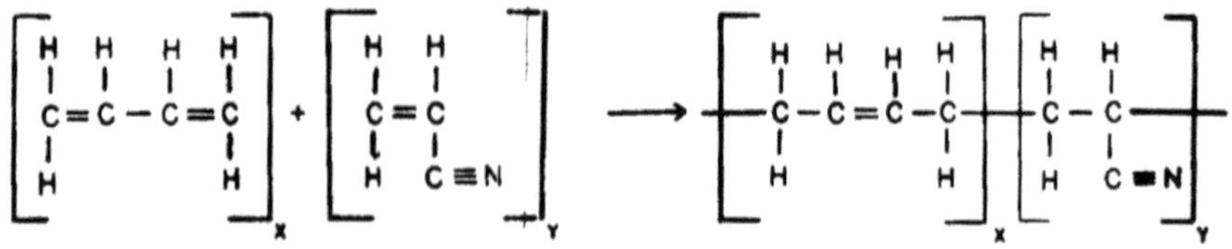

Figure 6.1 Copolymerization of butadiene and acrylonitrile to produce NBR.

Butadiene	At 30°C (86°F)	At 7°C (45°F)
trans-1,4	52%	Higher
cis-1,4	8%	Same
cis-1,2	7%	Lower
Acrylonitrile	33%	Same

These investigators observed that the cold polymerization results in increased *trans*-1,4-butadiene content at the expense of the 1,2-butadiene, giving a more linear structure with better processing, higher methyl ethyl ketone (MEK) solubility, and higher compound viscosity.

The properties of NBR vulcanizates depend greatly on the ACN content of the polymer. Two mutually incompatible requirements in a rubber are oil resistance and low temperature flexibility. As acrylonitrile content increases, the oil resistance increases. For example, tested under the same condition a 20% ACN content nitrile could show a volume increase in a common test medium, ASTM oil #3, of 59%; in a 50% ACN polymer it might be 7%. Furthermore, suppose the 50% ACN rubber had low temperature brittleness +3°F at 20% this might be as low as −69°F. Some other properties affected by the monomer ratio are processibility, cure rate, heat resistance, and resistance to permanent set. With increasing ACN content, processibility is easier, cure rate faster, and heat resistance better. However, resistance to permanent set is decreased. With increased viscosity, lower die swell and greater firmness in uncured parts can be expected.

Nitrile rubber is relatively expensive, and frequently attempts are made to obtain cheaper compounds by blending it with other polymers. Overall the results are spotty. Up to perhaps 15% natural rubber can be used to ease processing or increase building tack, but the rubbers are basically incompatible, and above such levels physical properties are impaired. More useful blends are made with SBR that is not oil-extended. Even up to 40% SBR (on the polymer content) the loss in tensile strength, elongation, and modulus is tolerable. Blends with NBR are particularly useful in the making of hard articles by blending NBR with finely divided phenol formaldehyde resins. These thermoplastic resins improve processing but give very hard compounds at room temperature.

B. Manufacture

Much information about the manufacture of nitrile rubber is proprietary, so it is difficult to give exact polymerization formulas and details such as times, temperatures, pressures, and purification procedures. Nitrile is made by emulsion polymerization like SBR; in fact some production facilities are interchangeable. There are hot and cold nitriles, depending on the temperature at which the rubber is made. That made at temperatures exceeding 30°C is called hot rubber, between 5 and 30°C we have cold rubber. NBR can be made by either a continuous or a batch process; the more popular grades are made continuously in polymerization lines of autoclaves. A general-type formula is given by Dunn and Blackshaw [4].

Butadiene	67.0
Acrylonitrile	33.0
Water	200.0
Emulsifier	3.5
Electrolytes	0.3
Catalyst	0.1
Activator	0.05
Modifier	0.5
Short stop	0.1
Stabilizer	1.25

Agents used to emulsify the monomers in water can be rosin or fatty acid soaps or such alkylaryl sulfonates as sodium dibutyl naphthalene sulfonate. Similar to that for SBR, emulsification is achieved by agitation, and an initiator is required to start the reaction. For hot polymerizations the activator system might be one given by Hofmann [5]:

Potassium persulfate	0.4
$FeSO_4 \cdot 7H_2O$	0.0005
Triethanolamine	0.4

Dunn and Blackshaw [6] note that the most frequently employed "redox" activation system is one containing a chelated ferrous salt and a sulfoxylate reducer, was developed by Polymer Corporation (now Polysar Rubber Division of Mobay Corporation) in 1949.

A modifier such as dodecyl mercaptan is used to control molecular weight and can be dissolved in the monomer butadiene before emulsification. Since modifiers are used up during polymerization, they are added incrementally during polymerization to ensure a constant chain length. Polymerization times rarely exceed 48 hr, and conversions from 60 to almost 100% are possible.

Polymerization is stopped by the addition of a short stop, ordinarily a water-soluble reducing agent. Butadiene monomer can be removed from the latex by lowering the pressure (butadiene boils under normal pressure at $-4.4°C$); acrylonitrile is stripped in a column with steam. Heat requirements are low, since acrylonitrile boils at 77.3°C. Acrylonitrile is poisonous and must be effectively removed if the rubber is to be used in contact with food products. Furthermore acrylonitrile has a disagreeable odor.

A stabilizer is added to the latex before drying and storage to protect it, since gel formation can occur during shelf aging. Commonly *tert*-butyl-*p*-cresol and trinonylphenylphosphite are used

Nitrile latices are most often coagulated and precipitated by adding salts such as NaCl or $CaCl_2$. For nitrile produced using soap as an emulsifier, it is important to remove the emulsifier because of swelling effects. This can be done by using dilute acids to hydrolyze the soap into free acids. The final step before drying is a thorough washing of the rubber crumb to remove precipitation agents, activators, emulsifiers, etc. This can be done by using filter drums or shaker screens with water sprays. Alternatively, the whole drying process can be accomplished by drying drums or belt dryers. After drying the crumbs are compacted into bales, which are then polyethylene-wrapped and packed for shipment.

C. Grading

The most common way to classify nitrile rubbers is by their bound acrylonitrile content followed by their viscosity. Equivalent grades can be determined by careful comparison of suppliers' literature. The bound acrylonitrile content, as mentioned earlier, goes from about 20% to 50% with the bulk in the 25–40% range. Mooney viscosity ordinarily ranges from 30 to 80, although Polysar markets a special grade, Krynac 34.140, which has a viscosity of 140. It is used as a viscosity modifier and for plasticizer masterbatches because it absorbs high loadings of plasticizers.

The test recipe developed by the ASTM for evaluating nitrile rubbers is that given in ASTM D3187–90 [7].

NBR	100.00
Zinc oxide	3.00
Sulfur, coated	1.50
Stearic acid	1.00
Oil furnace black	40.00
TBBS	0.70
	146.20

Except for the polymer and the sulfur, the ingredients in the formula are NBS standard reference materials; TBBS is *N-tert*-butyl-2-benzothiazolesulfenamide. Polymer producers probably would use the current industry reference black rather than the NBS material with little difference likely in the results. Sulfur is less soluble in nitrile than in natural, so that a surface-coated sulfur is recommended to get better dispersion. The coated grade contains 2% $MgCO_3$. Suppliers frequently give results based on the ASTM formula.

Besides the common NBR types, there are special grades available for greater versatility and specific applications. With practically worldwide production of this polymer, it is difficult to keep lists of such modifications current. Typical are two examples from North American suppliers. Copolymer Rubber and Chemical

Corporation market particulate black masterbatches under the trade name Nysynblak. One such masterbatch has 50 parts of N550 [a fast-extruding furnace (FEF) black], another 75 parts of N787 (SRF) black. The product is supplied as free-flowing pellets averaging 2–3 mm in diameter. Advantages of course are significantly less power consumption in mixing and shorter mixing times.

Similarly, Polysar Rubber Division of Mobay has available a masterbatch of NBR and plasticizer, with the trademarked designation Krynac 843. The masterbatch has 50 parts of diethylhexyl phthalate with 100 of a 34% acrylonitrile content polymer and can shorten mixing time and reduce handling of plasticizers.

D. Compounding

As noted before, NBR is used for its oil- and solvent-resistant properties. A grade is selected with the oil resistance that seems adequate for the product in service. Again, there are tradeoffs between high oil resistance and low temperature flexibility.

In essence nitrile is compounded much like natural rubber, although since it does not crystallize, reinforcing fillers are necessary to obtain reasonable tensile, tear, and abrasion levels. In black compounds reinforcement is proportional to the fineness of the black; blacks such as HAF give more than 3000 psi tensile strength at a 50-part loading with 25 parts of plasticizer. Blends of black are often used to fine-hone the qualities wanted. Nonblack fillers give the best heat resistance and also can be used with compounds that might be in contact with food products, where the amount of black used is restricted by FDA regulations. Fine precipitated silica is the most reinforcing of the nonblack fillers, but used in higher quantities it gives stocks that are hard and boardy.

Plasticizers are used in practically all nitrile compounds, first to aid processing and then to improve low temperature flexibility, resilience, and flexing, or to reduce hardness. Plasticizer levels will generally be between 5 and 50 parts on the polymer. Below 5 parts processing can be difficult; over 50 parts the material may bleed out and physical properties deteriorate unacceptably. Three types of plasticizer are usually used: organic esters, to get the best in low temperature flexibility; coal tar derivatives such as coumarone indene resins, to maintain tensile properties and improve building tack; and polymeric esters, to impart resistance to high temperature aging. The latter type of plasticizer has low volatility and is difficult to extract.

Nitrile rubbers require age resisters in addition to those they already have in order to give long service. Where staining is not a problem, amine-type antioxidants may be used. If staining is objectionable, then phosphite or hindered phenol antioxidants will serve better. If ozone protection is needed, waxes such as paraffin or ozokerite should be added. This is further discussed in Chapter 14.

Vulcanization of nitrile rubbers is usually accomplished using sulfur, accelerator, and zinc oxide and fatty acids as activators, although peroxides may be used in special cases. Vulcanization agents and procedures are dealt with in subsequent chapters, but some unique characteristics of nitrile curing and curative systems are noted here.

Nitrile rubbers cure relatively fast, with higher acrylonitrile content rubbers curing faster than lower content ones. The amount of sulfur used is based on the butadiene content and is generally lower than that used for natural rubber, ranging from 0.70 to 2.30. The solubility of sulfur is considerably less than in natural rubbers, which hinders its uniform dispersion. With poor dispersion some regions may be overcured, others undercured. For this reason sulfur is usually added early in the mix or as a masterbatch. Alternatively, special grades of sulfur may be used, such as those dispersed on a small percentage of carbon black or mineral filler.

Because of the relatively fast rate of vulcanization, nitrile is commonly compounded with only one accelerator, often of the sulfenamide class. However, if accelerators of the thiazole type are used, a second activating accelerator like tetramethylthiurammonosulfide might be used. A typical cure system for medium acrylonitrile level polymer could be 1.5 parts of benzothiazyl-2-diethyl-sulfenamide with 1.0 part of sulfur. Activating this vulcanizing system would be 5.0 parts of zinc oxide and 1.0 part of stearic acid.

Nitrile compounds used in seals where there are moving metal parts must have a low coefficient of friction. One way of achieving this is to add small quantities of graphite or molybdenum sulfide.

A new development in nitrile rubber has been the introduction of hydrogenated nitrile rubber (HNBR). This material can be made by dissolving the nitrile in solvent, hydrogenating in reactors, coagulating the resulting product, drying in an extruder, and packaging. The first commercial polymer of this kind was marketed by Nippon Zeon in 1984. The saturated ethylene chain part gives the rubber resistance to heat, ozone, and chemicals while maintaining oil resistance, and the butadiene section allows crosslinking.

Some demanding applications are being met by employing these hydrogenated nitrile rubbers. In oil well drilling HNBR is used for such parts as blowout preventers, pipe protectors, and well head seals, where the rubber must resist high temperatures, oil, hydrogen sulfide, steam, and explosive decompression. Another area is in the more critical automotive rubber parts, such as fuel hose tubes, which must be resistant to sour gasoline, and timing belts. Requirements for the latter items are particularly difficult to satisfy in that the compounds must have stable hardness, modulus, and dynamic properties over a wide temperature range, combined with good oil resistance. Properly compounded hydrogenated nitrile rubber can stand up to temperatures of 150°C. Presently there are two suppliers of such rubbers, Nippon Zeon and the Polysar Rubber Division of Mobay Corporation. These HNBRs usually have ACN contents in the 36–50% range; saturation will vary from 80 to more than 99%; and Mooney viscosity will be targeted between 55 and 85 (ML 1 + 4 @ 100°C). Other things being equal, as the nitrile becomes more saturated as the hydrogenation proceeds, ozone, chemical, and heat resistance all improve, although there is some decrease in dynamic properties. What could be a typical formula for an HNBR timing belt, given in a paper by Klingender and Bradford [8], is as follows.

Zetpol 2020	100
Zinc oxide	5
Stearic acid	1
N-770 (SRF)	40
Thiokol TP-95	5
Sulfur	0.5
TMTD	1.5
MBT	0.5
TMDQ	1
IPPD	1
	155.5

Curing conditions: 20 minutes a. 160°C.

Zetpol 2020, a product of Nippon Zeon, has 36% ACN, 10% unsaturation, and a Mooney viscosity of 71–85. The other additives are as follows: TMTD, tetramethylthiuramdisulfide; MBT, mercaptobenzothiazole; TMDQ, 2,2,4 trimethyl 1,2 dihydroquinoline; IPPD, N-isopropyl-N'-phenyl-*para*-phenylenediamine. In addition, we list several original properties

Original property	Value
Hardness, Shore A	70
100% Modulus, MPa (psi)	2.3 (330)
200% Modulus, MPa (psi)	5.4 (780)
300% Modulus, MPa (psi)	9.4 (1360)
Tensile strength, MPa (psi)	26.6 (3860)
Elongation, %	510

The low temperature properties of hydrogenated nitrile rubber, which were studied by Hayashi et al. [9], are different from those of NBR. The authors found that the low temperature flexibility of HNBR is strongly influenced by the degree of hydrogenation versus acrylonitrile content of the NBR. For sulfur-cured HNBR, di butoxyethoxyethyl)adipate (DBEEA), marketed by Thiokol as Thiokol TP-95, is one of the most recommended plasticizers. It shows the optimum balance between low temperature resistance and heat resistance at 125°C.

III. POLYSULFIDE RUBBER

A. Properties

The only polysulfide rubbers made commercially are those from the Thiokol Corporation, sold as the trademarked brands Thiokol FA, Thiokol ST, and Thiokol LP. The original polymer was discovered by J. C. Patrick in 1927, when he was trying to make ethylene glycol by hydrolysis of ethylene dichloride with various alkaline compounds.

Polysulfide rubbers are marketed in minuscule amounts compared to general-purpose polymers, but compounders should be acquainted with the products as they do have properties not matched by other polymers. Even before listing their superior properties, it should be noted that they have a disagreeable odor and a low tensile strength—up to 1500 psi. Molded compounds must be cooled before removal from the press, otherwise shallow depressions or craters will form on the surface.

Good properties of polysulfide rubbers are outstanding volume swell resistance to aliphatic and aromatic solvents, alcohols, ketones, and esters. Table 6.1 illustrates the resistance to volume swell after immersion of Thiokol's formulation 3000 FA (polysulfide rubber compound), whose principal components are as follows [10]:

Component	Parts by weight
FA 3000 polysulfide rubber	100.0
Zinc oxide	10.0
Stearic acid	0.5
MBTS	0.3
DPG	0.1
N774	60.0

Table 6.1 Volume Swell (%) After Various Periods of Time for Polysulfide Rubber Compound 3000 FA

Solvent	Immersion			
	1 day	1 month	6 months	1 year
Toluene	47	55	60	63
Ethyl acetate	20	19	19	16
Acetone	23	21	19	18
Carbon tetrachloride	30	37	39	40
Ethanol	3	3	3	3
Benzene	83	96	112	121

where MBTS is mercaptobenzothiazoledisulfide, DPG is diphenylguanidine, and N774 is a carbon black grade. MBTS has another commonly used name, benzothiazyldisulfide. For consistency and ease of recognition we will use the previous name mercaptobenzothiazoledisulfide as it relates so easily to the acronym.

The same publication [10] shows the relative resistance of polysulfide rubbers to medium acrylonitrile content nitrile, high acrylonitrile nitrile, and neoprene W. After 48 hr immersion at 80°F in SR-6 fuel, the polysulfide compound FA 3000 had only 10% swell compared to 25% for the high acrylonitrile nitrile, 60% for the medium, and 90 for a neoprene W compound. SR-6, a reference fuel containing 60% aromatics, is frequently used for testing solvent resistance. Thiokol FA is typical of polysulfide rubbers, and tests on it indicate what might be expected from other polymers of that type.

Other good properties of these rubbers are their low permeability to various solvents and to bases, as well as their low temperature flexibility and good aging resistance. Specific permeability through a 1/16" FA polysulfide sheet at 75°F (24°C) is only 0.005 for SR-6, 00.1 for carbon tetrachloride where the specific permeability is (oz) (in)^2/(24 hr.) (ft). Polysulfide rubber compounds have a service temperature range of -50 to $+250$°F. Low temperature flexibility is achieved without the addition of plasticizers.

B. Manufacture

Unlike the other synthetic rubbers covered, polysulfide rubber is made by a condensation reaction. Essentially this polymer is produced by the reaction of organic dihalides in a warm, agitated aqueous solution of sodium polysulfide in the presence of wetting and dispersing agents. The dispersed product is washed free of soluble salts and is coagulated by acids. The reaction can be represented by:

$$ClCH_2CH_2Cl + Na_2S_4 \rightarrow (CH_2CH_2S_4)n + 2NaCl$$

The organic halides used are ordinarily bis-(2-chloroethyl) formal and ethylene dichloride. Type FA is a linear polymer, presumably with hydroxyl groups at the end of the chain. These groups could be formed by hydrolysis of the organic dihalide during polymerization. Type ST has a branched structure when formed; a second treatment, which cleaves some of the disulfide bonds and results in thiol groups at the end of the chains rather than hydroxyl groups, imparts processibility.

C. Grades

Most compounding is done with the FA and ST polymers described above. The supplier's literature should be consulted for use of the liquid polymer types.

D. Compounding

FA rubber does not break down on milling to a workable plasticity. It is softened to a satisfactory level by using mercaptobenzothiazoledisulfide (MBTS). The process can be speeded up by the addition of diphenylguanidine (DPG). A starting ratio for plasticization would be 0.3 MBTS to 0.1 DPG. The amount of plasticization is dependent on the MBTS level, and wide swings in plasticity can occur with comparatively small changes. If the amounts are not weighed carefully and there is a small excess, a mushy, nonreclaimable stock on the mill will result. In determining the right amount to add, do not vary the MBTS by more than 0.05 part. The ST grade is millable as received.

For fillers, carbon blacks are used to reinforce and reduce the cost of the compound. Preferred black grades are thermal blacks as well as semireinforcing and fast-extruding furnace blacks. White fillers used include zinc oxide, lithopone, titanium dioxide, and blanc fixe. Fillers such as clays retard vulcanization. With FA, a stock containing 60 parts of a semireinforcing furnace black should have about 1200 psi tensile strength, 230% elongation, and a hardness of 68 Shore A. Softeners and plasticizers are rarely used in polysulfide stocks. Tackiness can be improved a bit by adding a low melting point coumarone indene resin. For curing, FA zinc oxide at 10 parts is about standard. ST would probably use about 5 parts of zinc peroxide and 1 part of calcium hydroxide. For lubricating the stock and to keep it from sticking to the mill, 1 part of stearic acid should be added, which also helps dispersion.

IV. RECYCLED RUBBER

Unlike the polymers described up to now, producers of recycled rubber seem to come and go as perhaps one reclaiming plant shuts down while a new company offering ground rubber debuts. Would be users can find current suppliers in the rubber trade press.

A. Reclaim

1. Properties

To avoid confusion, some clarifications are necessary in considering recycled materials. The term *recycled rubber* is an all-inclusive term that covers ground rubber, reprocessed synthetic rubber, hard rubber dust, die-cut punched parts (e.g., the cheap sandals, used in parts of the Far East, which are cut from tire sections), and reclaimed rubber. Here we are concerned only with reclaimed rubber and ground scrap rubber. Reclaimed rubber was defined by the rubber recycling division of the National Association of Recycling Industries in 1981 as follows:

Reclaimed Rubber is the product resulting when waste vulcanized scrap rubber is treated to produce a plastic material which can be easily processed, compounded and vulcanized with or without the addition of either natural or synthetic rubbers. It is recognized that the vulcanization process is not truly reversible; however an accepted definition for devulcanization is that it is a change in vulcanized rubber which results in a decreased resistance to deformation at ordinary temperatures.

Like virgin polymers, reclaimed rubber is sold in bale form. It is dark in color with a softer feel than the unprocessed polymer. Besides reclaimed natural, neoprene, and butyl rubbers, there are reclaims indicating the source, such as whole-tire reclaim and inner tube natural and inner tube butyl reclaim. Density varies from about 1.16 to 1.36 g/cm^3. The rubber hydrocarbon content (RHC) of reclaims varies considerably from about 40% to 60%. For compounding purposes the RHC is important, yet traditional methods of determining it now have serious errors that are due to the popular use of highly oil-extended SBR. A suggestion has been made that whole-tire reclaims (those most liable to error) be divided into three density classes and that the value for RHC be assumed for each class as follows [11]:

Density	Assumed RHC (%)
1.17–1.19	50
1.20–1.25	45
1.26–1.30	40

This proposal is believed to be in accordance with current practice.

During reclaiming, the molecular weight of the polymer present is significantly reduced from the original value. Chain scission occurs between carbon-to-carbon bonds. Chemical unsaturation is unchanged. Reclaiming agents such as dipentene or chemically unsaturated resin oils of petroleum origin may be added to improve processing, maintain tensile values, etc. Reclaim may be used as the sole polymer, but more often it is used in blends with virgin rubbers. Advantages of using reclaims in blends are as follows.

1. It makes lower cost compounds possible.
2. It shortens breakdown and mixing times.
3. Calendering and extrusion speeds often can be higher.
4. It maintains the (dimensional) stability of uncured stocks.
5. Because reclaimed rubber stocks have lower nerve than virgin polymer

stocks, there is lower swelling on extrusion and less shrinkage on calendering.

6. Reclaim cures relatively fast, so that acceleration dosage can be reduced.

2. *Manufacture*

Most reclaim is made from scrap tires, and the product is called whole-tire reclaim. Several processes have been developed for making reclaims of this and other types, such as butyl inner tube reclaim.

Irrespective of the process used, the first problem is to prepare the scrap rubber for reclaiming. With whole tires as the source, this means separating and disposing of the metal component (bead wire and belt wire) and the fiber content (tire cords). The primary equipment used to reduce the feed tires to a workable size scrap is the corrugated two-roll cracker mill. The rollers tear the tire up, and the heavy bead wire can be removed from the tire before the cracking operation or pulled out after the first pass. The bead wire can be salvaged as scrap metal. Passing the cracked rubber over magnetic separators serves to reduce remaining bead wire and belt wire. The product from the cracker mill is passed over vibrating screens, where oversized pieces are returned to the mill for further size reduction until a desire uniform size is obtained. A large portion of the fiber can be removed from this crumb by air separation technics. Continually passing the product over magnetic separators during all processing can almost eliminate ferrous metal fragments. Not only the metal part of the tire but also metallic items like nails embedded in the tread must be removed. The scrap rubber, now relatively free of fiber and metal, is ready for reclaiming. Several processes have been developed for this, both batch and continuous.

The simplest process—a batch process—is called the heater or pan process. The scrap is mixed with reclaiming oils and then placed in pans, which are rolled into a large, horizontal autoclave. In this vessel the scrap is heated with live steam at pressures ranging from 100 to 250 psi for 5–12 hr. After this treatment the cake from the pan is broken up, and reinforcing or processing agents are added if required. The resulting mix is passed through refining mills, strained to remove nonmagnetic metal and foreign material, and passed through a finishing and refining operation. The latter rolls are set only 0.003–0.005 in. apart, and any hard particles are moved at the side of the rolls as tailings. The product can then be sold as slabs, bales, pellets, or particles, depending on the buyer's preference. This pan process is really more suited to reclaiming butyl inner tubes than whole tires.

A second batch process has been called the dynamic dry digester process or dynamic devulcanization. The term "digester" stems from a reclaiming process, now obsolete largely for environmental reasons, in which reclaim was digested with water and salts such as NaOH to break down the cotton or rayon tire cord. In the superseding process, very finely divided crumbs with the necessary reclaiming oils are charged to a horizontal autoclave, which is fitted with an axial shaft to which paddles are attached. The rubber scrap is thoroughly agitated while being cooked with live steam at up to 375°F. With the high temperature and constant agitation, cooking cycles are short (usually 5 hr). Finishing operations are similar to those used for heater or pan process reclaim.

The only continuous reclaim process is the one called the reclaimator process. The scrap rubber is finely ground to approximately 30 mesh before being mixed in the blending system with the reclaiming oils. Then it is charged to the reclaimator, a machine, like an extruder, which is jacketed with several zones so that varying temperatures can be achieved; the discharge end is cooled. The screw is specially designed and has only a small clearance with the inside of the barrel. The end result is that the scrap is constantly and intensively worked on under high heat. Temperatures between 400 and 500°F are reached in the charge, and this along with the working gives a through-pass time in minutes compared to hours. After emerging at the discharge end, the reclaim is cooled and finished as other reclaims are. This process yields a very uniform product, production rates are high, and there is no environmental pollution.

3. Grading

There are not well-standardized grade limits for reclaim. Much reclaiming is done, for example, on a contract basis. The Rubber Reclaimers Association test recipe for tire and natural rubber reclaims is as follows:

Reclaimed rubber	200.0
Zinc oxide	5.0
Stearic acid	2.0
2-Mercaptobenzothiazole	0.5
Diphenylguanidine	0.2
Sulfur	3.0
	210.7

Cures are made at 287°F. The reclaim is assumed to contain 50% RHC.

4. Compounding

Before using any reclaim, the compounder should obtain from the supplier a sample (not freshly made) and examine it for hard lumps and flakes of metal by tearing the sample apart in several places. (Since the material is built up from fine sheets, separation is not difficult.) The rate of cure should be checked out, and staining behavior should be determined, if this is important. Reclaimed rubber when only 2 or 3 days old has a rheological behavior different from the same material 7–10 days old. Accordingly, for uniformity in production processing like extrusion, the more stable material should be used.

If an all-reclaim compound is desired, the test formula given above could be used as a base with the addition of blacks, nonblack fillers, and oils to achieve the cost and quality level wanted.

Usually reclaim is used in blends for processing reasons and to reduce costs. Based on an estimated RHC of 100, loading materials would probably vary from 100 to 150 parts. For example, a steam hose cover might have 60 parts of natural rubber plus 80 parts of whole-tire reclaim (50% RHC) and 100 parts of medium thermal black. A black automotive mat might be compounded with 100 parts of whole-tire reclaim and 35 parts of SBR 1712, 80 parts of clay, 40 parts of hard rubber dust, and 15 parts of process oil.

B. Ground Rubber

The distinction between conventional reclaimed rubber and ground rubber may be narrowing. This section deals with ground rubber, which has not been treated to softening as happens in many reclaiming processes. A development of the last few years has been the introduction of ground rubbers having very fine particles. In grinding scrap rubber by mechanical methods, the product is usually in the − 40 mesh range. Finer grinds would be preferable, but the difficulty is that in greater reductions the rubber suffers heat degradation and sticks together. Finer particle rubbers, passing an 80 mesh screen, can be made by cryogenic grinding or new wet grinding techniques.

In sieve classification, *mesh* is the number of openings per linear inch. A 40 mesh screen then has openings 420 μm on a side, an 80 mesh screen 177 μm on a side. Cryogenically grounds rubbers passing an 80 mesh screen are more expensive than the wet-ground material and have a narrower band of particle sizes at a higher mean particle size. The mean particle size for cryogenic ground rubber might be 163 μm, that for a wet grind product like UltraFine™ Grind 74 microns.

The use of these ground rubbers passing an 80 mesh screen (− 80) does provide some benefits in compounding over other fillers. They have low specific gravity (which lowers cost on a volume basis) and high resilience, and they contain the antioxidants, antiozonants, and UV inhibitors of the original compound. Since scrap tires provide the usual raw feed for ground rubber, most fine grinds are SBR or NR/SBR based, but such ground rubbers are available in butyl, EPDM, neoprene, nitrile, and natural rubber types.

A prominent U.S. manufacturer of such finely ground rubbers is Rouse Rubber Industries, and the development and applications of these compounds are reviewed in a paper by M. W. Rouse [12].

One Rouse product that can be used in tires, molded goods, and footwear is GF 80 Ultra Fine (Powder). The chemical properties are as follows:

Moisture–volatile matter	1%
Fixed carbon	28%

Carbon black	27–33%
Hydrocarbon content	42%
Acetone extract	10–20%
Ash	5%
Total sulfur	1.0–2.0%
Free sulfur	Trace
pH	6.0–7.0

Such an analysis reflects the composition of present-day tire compounds, especially treads.

Physical properties of the same material include the following:

Specific gravity	1.12–1.15
Surface area, m^2/g	2.0
Average particle size	200 mesh
Screen analysis	
passing 60 mesh	100%
passing 80 mesh	90%
passing 120 mesh	85%
passing 200 mesh	50%

Such a rubber could be used as the sole polymer in matting or molded goods, with the addition of fillers, oil, and curatives to obtain lower costs and a tighter cure.

One of the most critical tire compounds is the tread component. Rouse [12] gives some interesting data comparing the properties of a control sample of a radial tire tread stock and those when a 10-part substitution with ground rubber passing an 80 mesh screen was made. That is, the experimental stock used 90 parts of control stock and 10 parts of ground rubber (GF UltraFine (Powder)). Test results at the same cure time were as follows:

Property	Control	10% Substitute
300% Modulus, MPa (psi)	8.1 (1170)	8.6 (1250)
Tensile strength, MPa (psi)	15.9 (2310)	15.3 (2220)
Elongation, %	520	495
Durometer, Shore A	71	71
Abrasion rating	94	92

The reduction in values seems small considering the amount of substitution. One of the basic qualities required in a tire tread is good wear or abrasion resistance. Given the variability inherent in abrasion tests, it was not judged significant that the ground rubber substituted compound was 2 points lower in the abrasion resistance rating.

Table 6.2 Selection Guide for Elastomers[a]

Property	Natural (NR)	Styrene-butadiene (SBR)	Polybutadiene (BR)	Neoprene (CR)	Butyl (IIR)	Nitrile (NBR)	Ethylene propylene (EPM, EPDM)
Bondability to metals	E	E	E	E	G	E	G
Density, g/cm^3	0.93	0.93	0.92	1.23–1.25	0.92	0.88–0.90	0.86
Tensile strength, psi (reinforced, max)	4500	3600	2500	4000	3100	3800	3000
Elongation at break, max	550	450	475	600	750	700	500
Durometer, Shore A range	30–90	40–90	40–90	40–95	35–85	50–75	30–85
Resilience, 73°F	E	G	E	G	F	F	G
Low temperature flexibility	E	G	E	F	G	F	G
Resistance to abrasion	E	E	E	G	F	F	G
Resistance to tearing	E	F	G	G	G	P	F
Atmospheric aging (sunlight)	G	G	G	E	E	G	E
Resistance to heat aging	F	F	F	G	G	G	E
Resistance to oxidation	G	G	G	E	E	F	E
Resistance to ozone	F	F	P	G	E	F	E
Resistance to water absorption	E	G	G	G	E	G	E
Resistance to dilute acids	E	G	G	G	E	F	G
Resistance to concentrated acids	F	F	F	G	E	P	G
Resistance to alkalis	G	G	G	G	E	F	G
Resistance to gas/ oil/grease	P	P	P	G	P	E	P
Resistance to animal/ vegetable oils	P	P	F	G	E	G	F

[a]E, excellent; G, Good; F, Fair; P, Poor

Table 6.2 can serve as a guide to selecting the best elastomer for a given application.

REFERENCES

1. Dunn, J. R., and Blackshaw, G. C., *NBR: Chemistry and Markets*, a contribution by Polysar Limited to the Energy Rubber Group Educational Symposium, Dallas, TX, Sept. 24, 1985, p. 11.
2. Hofmann, W., in *Kautschuk Handbuch*, Vol. 1, S. Bostrom, ed., Verlag Berliner Union, Stuttgart, 1950, p. 380.
3. Dunn, J. R., and Blackshaw, G. C., *NBR: Chemistry and Markets*, a contribution by Polysar Limited to the Energy Rubber Group Educational Symposium, Dallas, TX, Sept. 24, 1985, p. 9.
4. Dunn, J. R., and Blackshaw G. C., *NBR Chemistry and Markets*, a contribution by Polysar Limited to the Energy Rubber Group Educational Symposium, Dallas, TX, Sept. 24, 1985, p. 4.
5. Hofmann, W., Nitrile Rubber, A Rubber Review for 1963, *Rubber Chem. Technol.*, *37* (2), 89 (1963).
6. Dunn, J. R., and Blackshaw, G. C., *NBR Chemistry and Markets*, a contribution by Polysar Limited to the Energy Rubber Group Educational Symposium, Dallas, TX, Sept. 24, 1985, p. 5.
7. *1984 Annual Book of ASTM Standards*, Vol. 09.01, p. 512.
8. Klingender, R. C., and Bradford. W. G., HNBR Exhibits Dynamic Properties Well Suited for Timing Belt Use; *Elastomerics*, *123* (8), 10 (1991).
9. Hayashi, S., Sakakida, H., Oyama, M., and Nakagawa, T., Low Temperature Properties of Hydrogenated Nitrile Rubber (HNBR), *Rubber Chem. Technol.*, *64* (4), 534 (1991).
10. Technical Bulletin, *FA Polysulfide Rubber*, Thiokol Corp. 2475 Washington Blvd, Ogden, UT. 89401.
11. Smith, F. G *Vanderbilt Rubber Handbook*, 1978, R. T. Vanderbilt Co., Norwalk, CT, p. 325.
12. Rouse, M. W. The Development and Application of Superfine Tire Powders for Rubber Compounding, presented at The Tire Industry Conference, Oct. 23–24 1991, Greenville, SC.

7

Vulcanizing Agents

I. INTRODUCTION

The ASTM definition for vulcanization reads as follows: "An irreversible process during which a rubber compound through a change in its chemical structure (for example, cross-linking) becomes less plastic and more resistant to swelling by organic liquids while elastic properties are conferred, improved, or extended over a greater range of temperature." This change can be brought about by a variety of agents including irradiation. Since the last method requires expensive equipment and works best on thin rubber sections, this chapter deals only with the conventional vulcanization materials. Again the terms "curing" and "vulcanizing" are used interchangeably.

There are four curing agents or systems in common use. They are:

1. Sulfur systems
2. Peroxides
3. Urethane crosslinkers
4. Metallic oxides (used in vulcanizing neoprene only)

By far the most common vulcanizing methods are those dependent on sulfur.

Before describing vulcanizing agents, it will be useful to review common vulcanization methods and their applications. In several instances the curing methods used or available restrict or modify the use of certain vulcanizing agents.

Almost all rubber products are vulcanized by one or more of six methods. These are:

1. Press curing (including injection and transfer molding)
2. Open steam curing
3. Dry heat curing
4. Lead press curing
5. Fluidized-bed curing
6. Salt bath curing

Press curing is used for items as diverse as hot water bottles, rubber mats, belts, and tires. Tire molding is a very specialized and complex case and need not be detailed here.

Belts, like conveyor belts, are considered to be molded because metal bars are placed on both sides of the uncured belt in the press to confine the edges and maintain the proper width. Compression molding is the simplest molding process. The uncured rubber blank is placed directly in the mold before the mold is closed and placed in the press. In transfer molding the uncured rubber is moved from a reservoir into the mold cavities while the mold is closed and in the press.

Injection molding is somewhat similar to transfer molding in that the rubber compound is pressed into a closed cavity through sprues. It differs in that a measured volume of compound is forced into the cavities through runners by an injection head. The molds are clamped into presses, which are assembled in clusters, with the result that there may be 10 molds for a single injection machine.

Within broad limits any vulcanizing agent can be used in press cures. Some press operators may find the odor of dicumyl peroxide, a popular peroxide curing agent, objectionable. Press curing is usually done at 275–395°F. The higher temperatures are often obtained by using hot oil rather than saturated steam in the press plates.

It is often desired to determine equivalent cure times for different temperatures. As an approximation, the general rule that chemical reaction rates double for each 10°C or 18°F rise can be used. Accordingly, a molded article that cures satisfactorily in 40 min at 287°F should reach the same state of cure in 20 min at 305°F. It must be remembered that this is only an approximation and is most useful for articles ¼ in. thick or less.

Open steam curing is used for hose, wire and cable, expansion joints, and extrusions such as small tubing. The equipment usually consists of a vertical or horizontal autoclave, jacketed to avoid condensation in the autoclave, or possibly insulated. Provision is made for introducing steam, and valves are arranged so that the air trapped inside at the start of the cure is bled off to the atmosphere. Extrusions such as tubing may be coiled onto a bed of soapstone in pans before being placed in the autoclave, the soapstone giving support to the extrusion before firming up as it cures.

Open steam curing is used to vulcanize hand-built hose, usually made in 50-ft lengths. The hose, made on a hollow metal mandrel, is spirally wrapped with tape before going to supports inside the vulcanizer. The wrapping prevents sagging during the initial part of the cure. A curing process known as continuous vulcanization (CV) is used in making insulated wire and cable. The raw wire or cable emerging from the insulating tube is pulled into a jacketed tube, where open steam at pressures up to 200 psi cure the compound in a matter of seconds. The tube length may exceed 200 ft. Open steam methods like this do not allow the use of urethane crosslinkers for reasons more fully described later. Peroxide cures are used for ethylene propylene rubber insulations.

Footwear and coated fabrics are often vulcanized in dried, heated air. In using heated air, temperatures above 300°F are rather hard to maintain in conventional enclosures; accordingly, vulcanizing agents and cure systems that can function well at lower temperatures are used. Again peroxide cures cannot be used, since oxygen disrupts the normal mechanism of peroxide vulcanization and leads to chain scission and polymer breakdown.

Lead press curing is used for some kinds of electrical cable and hose. In this curing method, the hose or insulated wire, after passing through an orifice, is immediately encased in a lead sheath whose thickness can be varied. The sheath serves as a mold for the finished product. A cylindrical slug of lead, weighing perhaps 1100 lb and kept at an elevated temperature for easier extrusion, is dropped into a cavity in the press, and a ram pushes the metal around the rubber article. When curing is complete, the lead is stripped off, cut into small pieces, remelted, and reused.

The advantages of lead press curing include the efficiency of the method, a smooth outside appearance, and the length of product that can be vulcanized. By coiling the lead-sheathed product and then vulcanizing, long lengths, say 600 ft of hose, can be obtained.

One way of curing lead press hose is to load the leaded lengths into an autoclave and make liquid-tight passages between the lengths. Saturated steam is then introduced on the outside and circulating superheated water inside the hose lengths. The internal pressure assures a dense, well-bonded hose with a smooth appearance when the lead sheath is peeled off. Since the common vulcanizing agent—sulfur—reacts with the lead to form dark lead sulfide, light colors are difficult to make. Again, if water is present in the hose tube during vulcanization, urethane crosslinkers cannot be used as vulcanizing agents.

Two relatively recent curing methods are liquid salt bath curing and fluidized-bed curing. Both methods are used for curing extrusions. In the first case the extrudate is rapidly passed under the surface of a long trough containing fused inorganic salts at a temperature between 350 and 400°F. Curing is accomplished in a matter of seconds, and the curative system must be chosen to accomplish this. It would be most impractical, for example, to rely on sulfur alone as the vulcanizing agent.

Frequently a more serious problem is the content of volatiles. At such temperatures, compounding ingredients like fillers with inadvertently high moisture levels (perhaps because of high humidity storage) will cause porosity as the water is quickly volatilized. Besides fillers with appropriately low moisture contents, desiccants are frequently added to the mix to avoid this trouble. Plasticizers with low boiling constituents can also cause porosity.

In fluidized-bed curing such products as automotive sponge stripping are cured by passing through a heated chamber containing small glass spheres called ballotini, which are constantly in motion and support the extrudate. Again peroxide crosslinkers cannot be used because of the oxygen present.

Before considering specific vulcanizing agents, two aspects of the curing process should be touched on. A common problem in rubber goods factories is scorch—the tendency of a compound to start curing before it is in the right form and the right place to be cured. There is a vicious circle here: for cost reasons, the quicker a compound can be cured, and the lower the temperature, the better; yet compounding to this aim alone means that more stock will be rejected because of scorch. Apart from careful compounding, scorch is lessened by mixing, storing, and processing the stock at the lowest temperature possible (yet breaking down and dispersing ingredients satisfactorily). Masterbatches of sulfur and accelerators added at the end of the mixing cycle help to keep heat history to a minimum. Stocks should be cooled quickly after mixing and not held for long times on warm-up mills prior to extruding or calendering.

Another point to be remembered is that few rubber goods items are fully cured when made. By "fully cured" is meant that state at which hardness is at a maximum, modulus is fully developed, and further curing might result in reversion (e.g., with natural rubber stocks). The reason for not going to a full cure is that rubber properties develop at different rates as curing progresses. For example, resistance to tear and abrasion usually peaks before resistance to tensile strength. On the other hand, swelling resistance is greatest when a compound is fully cured. So if top tear resistance is wanted, the best product cure might be an undercure. Again the curing may be finished in service, giving a longer life. For example, steam hose tubes are made with some undercure because curing will be completed in service.

II. SULFUR CURING SYSTEMS

Methods for curing with sulfur can be broken down into four different systems:

1. Sulfur alone
2. Conventional sulfur and accelerators
3. Low sulfur and accelerators
4. Sulfur donor systems

Methods 3 and 4 are called efficient vulcanizing systems because more mono-sulfidic than polysulfidic bonds are in the crosslinks between chains. This molecular arrangement makes for better heat resistance in the vulcanizates.

Rubbers like NR and SBR can be vulcanized with sulfur as the only curing agent. Commercially such products are usually called hard rubber items, such as combs, battery cases, and tank and pipe linings to resist chemicals. For these products over 25 parts of sulfur usually is used, but even for these products modern practice is to use a small amount of an accelerator like diphenyl guanidine (DPG). Many of the products once made from hard rubber are now being made from polyethylene and polypropylene. Current practice in hard rubber manufacture is reviewed by Cooper [1].

Two forms of sulfur are available to the compounder: regular and insoluble. Ordinary sulfur is crystalline and rhombic. This type is somewhat soluble in polymers, about 1 part per 100 of natural rubber at room temperature to about 5 parts at 75°C. If an NR compound has 3 parts of ordinary sulfur, the sulfur will dissolve during the hot mixing process. On cooling the sulfur which has migrated to the surface will crystallize, changing the appearance of the surface; this is called sulfur bloom. Bloom inhibits the bonding of uncured rubber pieces when building up products and must be removed by a solvent rub. Sulfur bloom can also occur on vulcanizates. This might happen when curing is incomplete and there is much unreacted sulfur. Here the disadvantage is largely cosmetic, since the appearance of the product is spoiled. All blooms are not sulfur blooms. For example, 5 parts of a paraffinic oil in a neoprene compound will result in exudation from the vulcanizate due to the solvent resistance of the polymer. Blooms on cured rubber are most often due to the vulcanization accelerator system.

Relief from sulfur bloom can often be had by using insoluble sulfur. This is an amorphous, polymeric sulfur with greatly reduced solubility in elastomers. Such sulfurs can contain so much oil (used to help dispersion) that sulfur dosage in the compound must be recalculated. Insoluble sulfur does tend to revert back to the rhombic form as the temperature is increased in mixing. At least one brand of insoluble sulfur has been stabilized to retard this reversion.

If sulfur bloom still exists even with a change to insoluble sulfur, one more step may be helpful. The insoluble sulfur can be dispersed in a rubber master-batch and added at the end of the mixing cycle. In this way the heat history of the sulfur is at a minimum and the tendency to revert to the rhombic form is lessened. Such dispersions are available commercially.

Two points are noted concerning the use of sulfur in compounding. Irrespective of the type used, poor dispersion can cause serious problems in vulcanizates. Frequently the fault is not due to an inappropriate mixing cycle but rather lumpy sulfur. In such cases the remedy is simply to "scalp" the sulfur: that is, screen it before adding it to the mix.

The second item has to do with the density of vulcanizates containing sulfur. Sulfur has a density of 2.06. In vulcanized rubber there is a reduction in total volume of the ingredients due to the crosslinking reaction between sulfur and polymer. Work done by Scott [2] at the National Bureau of Standards indicates that if a fictitious value of 6 is used for the density of sulfur in determining the vulcanizate density, this reduction is compensated for in normal sulfur level compounds. Except for hard rubber compounds, sulfur is rarely used at dosages exceeding 3.5 parts per hundred of polymer.

Compared to sulfurless vulcanizing, the use of sulfur brings some general advantages. Unlike peroxides, for example, sulfur does not react very much with ingredients in the compound other than the accelerators and activators. With a proper choice of the last two components, vulcanization with sulfur can be more controllable with regard to time and temperature. Contrasted to non-sulfur-cured vulcanizates, those cured with sulfur generally have higher tensile strength, tear resistance, and resistance to flex cracking. On the down side, sulfur-cured vulcanizates bloom from uncured or cured stock and have poor compression and elongation set at elevated temperatures, as well as poor resistance to aging.

The use of sulfur with conventional acceleration, low sulfur, and sulfur donor systems is discussed in Chapter 8 (Accelerators) because of the interactions involved.

III.   PEROXIDE CURING SYSTEMS

In the early 1950s the use of dialkyl peroxides to crosslink polymers was introduced. This method has some advantages over sulfur curing systems. Peroxides can crosslink both saturated and unsaturated polymers. The vulcanizates have better heat-aging properties, lower compression set, and good low temperature flexibility. In white and transparent vulcanizates there is freedom from color opacity and resistance to UV light discoloration. There are also some disadvantages. Peroxides are rather hazardous chemicals and require more attention to safe storage and handling procedures than many other compounding ingredients. Certain peroxides produce unpleasant odors in curing and in the vulcanizate. They react more readily with other compounding ingredients than conventional sulfur curing systems do, which restricts the use of antioxidants, for example.

Peroxides are not suitable for curing in the presence of oxygen, such as hot air cures. The reason is that a radical transfer from the peroxide to the polymer chain can be oxidized. This means that a hydroperoxide can form readily, and when this is thermally decomposed it leads to polymer degradation. Finally, since butyl rubber is degraded in the presence of decomposing peroxides, this method of cure cannot be used for IIR.

A. Physical Properties

Although diacyl peroxides and peroxy esters have been used in crosslinking, dialkyl peroxides are most frequently used. Typical members of this class are 2,5-dimethyl-2,5-di(*t*-butyl-peroxy) hexane, di-*t*-butyl peroxide, and dicumyl peroxide. The most popular is dicumyl peroxide. In their refined form these accelerators are liquids or solids, 90–99% pure. Because of their reactivity, they are frequently distributed on an inert carrier like calcium carbonate at a 40–50% concentration and used in this form. Some can also be obtained as dispersions in a polymer.

Dialkyl peroxides should be stored at 100°F or less to avoid decomposition. They burn actively and are difficult to extinguish, so they should be kept away from heat sources. Static electricity should be reduced by grounding when these agents are handled.

B. Peroxide Curing Process

Homolytic decomposition of the peroxide by heat produces two free radicals.

$$ROOR \rightarrow 2RO \cdot$$

For dicumyl peroxide two alkoxy radicals ($RO \cdot$) are produced, as shown in Figure 7.1.

Diacyl peroxides give acyloxyl radicals ($R{-}\overset{O}{\underset{}{C}}{-}O \cdot$), peroxy esters give both alkoxy and acyloxyl radicals. Both tertiary alkoxy and tertiary acyloxyl radicals can undergo still more decomposition by β-scission or decarboxylation to give alkyl radicals. All these radicals can start a crosslinking reaction.

In certain unsaturated polymers, like natural rubber, the free radical can react at the double bond to form a polymer radical. This in turn combines with additional polymers to give multiple crosslink bonding (Fig. 7.2).

If the polymer has an abstractable hydrogen, an R—H group is formed with the crosslinked polymer (Fig. 7.3).

Certain side reactions occur with peroxides that must be recognized and allowed for. For example, butyl rubber cannot be crosslinked by peroxides because tertiary carbon radicals suffer a β-scission reaction, which leads to

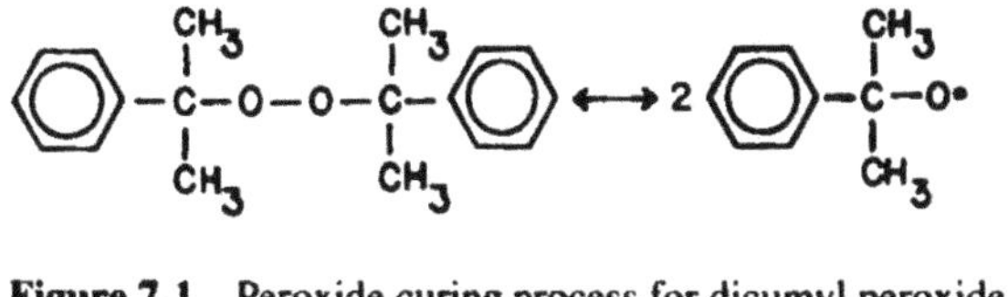

Figure 7.1 Peroxide curing process for dicumyl peroxide produces two alkoxy radicals.

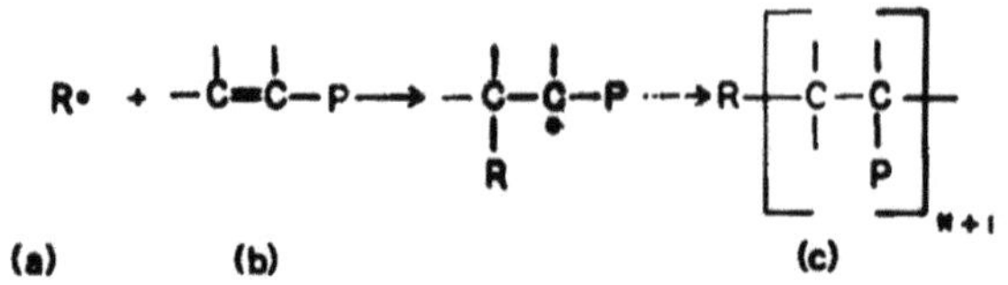

Figure 7.2 Multiple crosslink bonding in polymer radicals: (a) radical, (b) unsaturated polymer, (c) crosslinked polymer.

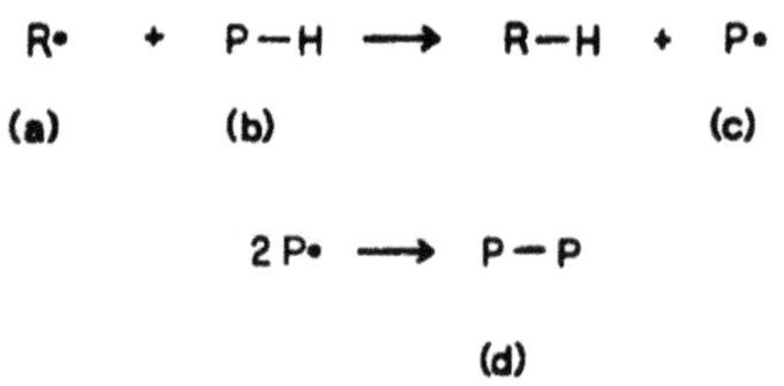

R•   +   P—H ⟶   R—H   +   P•

(a)   (b)   (c)

2 P• ⟶   P—P

(d)

Figure 7.3 Formation of a crosslinked polymer: (a) radical, (b) polymer, (c) polymer radical, (d) crosslinked polymer.

degradation of the polymer. More important is the possibility of competing reactions. Many compounding ingredients, such as oils and antioxidants, react with the radicals formed on peroxide decomposition, and these reactions dilute the crosslinking effect. As discussed in more detail later, care must be taken to limit such interfering materials.

C.   Using Peroxides as Vulcanizing Agents

NR, SBR, BR, EPDM. NBR and CR can be crosslinked by peroxides; IIR cannot. Mixtures of the first five rubbers can be crosslinked as well, although the state of crosslinking will depend on the solubility of the peroxide in the component polymers and the number of reactive groups and abstractable hydrogens in each polymer. This chapter considers only the most popular peroxides (i.e., the dialkyl peroxides). With this group, vulcanization is usually carried out at 300–450°F. The efficiency of a particular peroxide is measured by the number of crosslinks produced per mole of per-oxygen linkage. The measurement is made indirectly by determining the torque produced at varying curing temperatures in an instrument like the Monsanto oscillating disk rheometer. The higher the torque obtained, the higher the number of crosslinks created.

In optimizing conditions for mixing, bin storage life of the compound, and curing, it is useful to know the half-life of the peroxide chosen. Half-life, is the time by which half the peroxide is decomposed at the temperature chosen, and values given in suppliers' literature can vary with the conditions of measurement.

At any given temperature, 6 half-lives is considered to represent complete decomposition of the peroxide for practical purposes.

Few specific rules can be given here regarding the peroxide concentrations, curing temperatures, and vulcanizing time that should be used. Such decisions depend on the polymer or polymer blends involved; the fillers, oils, plasticizers, and age registers used; and the curing equipment available.

Some general considerations can be offered. Peroxide concentrations in commercial compounds usually do not exceed 2.5 parts per hundred. As concentration increases, modulus and elongation decrease; tensile strength has a tendency to peak and then diminish. Customary curing temperatures are between 320 and 380°F for the majority of compounds. Studies have indicated that for each 10°C (18°F) decrease in temperature, cure rate decreases by a factor of approximately 2.5 in the temperature range of 257°F (125°C) to 360°F (182°C) [3]. Increasing the cure time at constant temperature will reduce the compression set, often a desirable change, but perhaps at the cost of unacceptably lower elongation and higher modulus.

The effect of other compounding ingredients is more marked for peroxide systems than for all other vulcanization systems. Acidic carbon blacks, like channel blacks, slow the cure rate compared to sulfur vulcanization. Neutral or alkaline mineral fillers like fine particle silica or ground limestone have little effect. Particular care must be taken in the choice of oils and plasticizers, since free radicals from the peroxide can abstract hydrogen from them. Paraffinic oils appear to be the most resistant to this, aromatic oils the least. Similarly, there are reactions between peroxides and antioxidants and antiozonants. The latter should not be used; quinoline-type antioxidants have the least effect on peroxide cures. Some ozone protection can be obtained by using paraffin wax in the compound.

Overall, peroxide curing agents usually are not used with NR, SBR, BR, CR or reclaim. They can be effectively used with NBR, especially hydrogena NBR (HNBR) and highly saturated polymers like EPDM.

IV. URETHANE CROSSLINKERS

A. Introduction

One of the newest vulcanizing agents is a urethane crosslinker [4], discovered by scientists at the Malaysian Rubber Producers Research Association (MRPRA) in England and developed by Hughson Chemicals, Lord Corp., in the United States. Presently marketed by Rubber Consultants in England, by far the most popular type is sold under the tradename Novor 924. MRPRA applied the term "urethane" to these materials because of their analogy with regular urethane compounds, and Novor 924 is an adduct from a substituted nitrosophenol and toluene diisocynate. It is a dustless, free-flowing powder, 75% active ingredient and 25% naphthenic oil. The nature of its components makes it an expensive vulcanizing agent, and increased volume usage probably would not reduce the cost greatly. Because of the unique properties it imparts, it may be economical in certain cases, especially when used with a sulfur vulcanizing system. Recently it has been used with natural rubber, although there are references to its use with SBR [5] and neoprene [6].

Urethane crosslinkers were developed to improve the resistance to reversion and aging of natural rubber compounds. Since they have been in use for a relatively short time, their reactivity and their functioning have not been fully determined.

As this is written production and sales of Novor 924 is being replaced by Novor 950. This is being done for health and safety reasons in manufacturing. Novor 924 used toluene diisocyanate (TDI) and can emit diisocyanate in minute quantities. Novor 950 is made using methylene bis (4 phenylisocyanate) (MDI) with its much lower volatility meets health and safety regulations more easily. The producers state that Novor 950 will give the same results as Novor 924. In using Novor 950 in natural rubber compounds a weight for weight replacement of Novor 924 in any compound should be made.

B. Crosslinking Mechanism

The core reactions involved in using a urethane crosslinker like Novor 924 are shown in Figure 7.4.

The crosslinker is an adduct of a *p*-nitrosophenol and a diisocyanate, and the number of such adducts that can be formed is considerable. The adduct is a quinone oxime urethane as shown at the top of Figure 7.4. At vulcanizing temperatures the adduct decomposes, and the *p*-nitrosophenol units react with the rubber chain to give it pendant aminophenol groups. In turn the aminophenol groups are joined by the released diisocyanate to give urea-type crosslinks. It is the type of crosslinks which gives the rubber its heat-aging resistance.

C. Using Urethane Crosslinkers

Urethane crosslinkers are not used alone but always with a catalyst, such as zinc dimethyldithiocarbamate, which acts to increase the formation of pendant groups. In addition, a drying agent like calcium oxide is used to absorb moisture present in the rubber compound, which might hydrolyze free isocyanate, producing carbon dioxide and causing porosity in the vulcanizate. In most cases, for reasons of economy, urethane crosslinkers are used in conjunction with sulfur curing systems. Happily the two systems are synergistic, which further reduces the cost.

With the use of covulcanizing systems, natural rubber compounds can be devised that are resistant to reversion and heat aging and have improved fatigue strength after aging. Other advantages include extending the temperature range

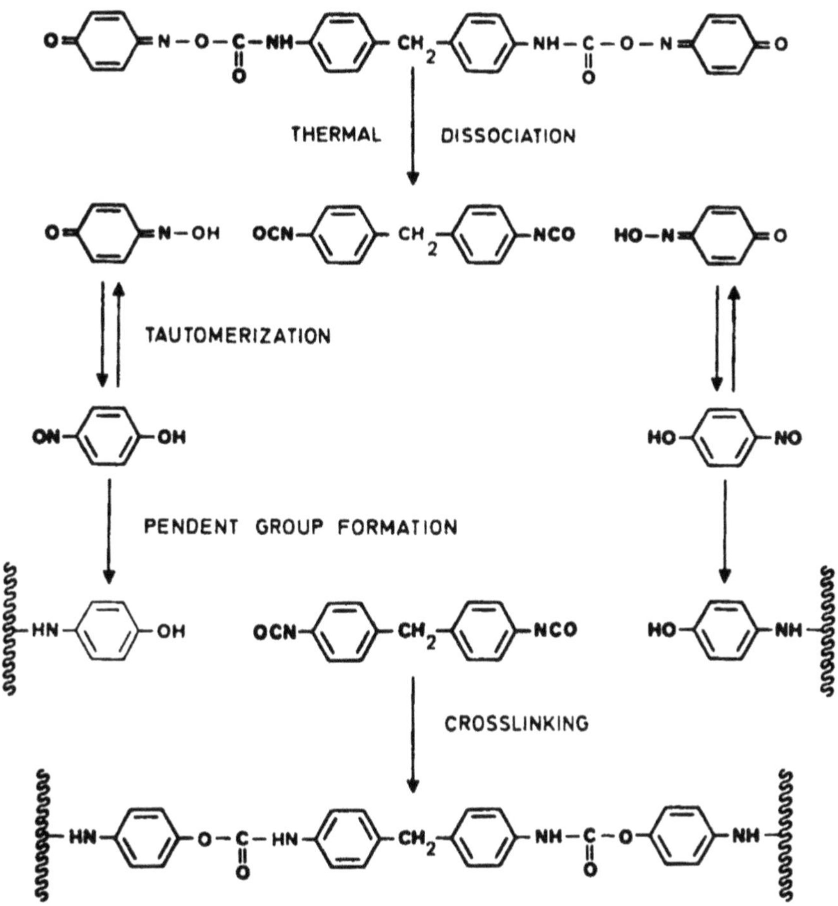

Figure 7.4 Crosslinking with Novor curing agent. (Courtesy of Plastics and Rubber Institute (Malaysian Section)).

for curing natural rubber and increased hardness of a compound compared to that from equivalent sulfur-cured vulcanizates. These advantages have resulted in Novor compounds being used for curing bags, rollers, engine mounts, and solid tires.

The compounding and handling of stocks mixed with urethane crosslinkers is distinctly different from those using conventional curing agents. As mentioned earlier, calcium oxide is usually added to absorb moisture, which otherwise would hydrolyze the isocyanate. In addition, fillers with high moisture content should be avoided, stocks should be stored in a dry place, and immersion in cooling tanks limited or avoided. In some instances wrapping the stock tightly in polyethylene film may be necessary.

Typical curing systems given by the supplier [7] are as follows:

100% B Novor Cure

Novor 924	6.7/pph (equivalent to 5.0/phr active ingredient)
ZDMC[a]	2.0/pph
Calcium oxide	5.0/pph

Novor/Sulfur Mixed Systems

An 80:20 ratio is typical.

Ratio	Novor 924	TMTM[b]	Sulfur	Santocure NS[c]
Novor/sulfur 80/20	4.2	1.3	0.4	0.08

[a]Zinc dimethyldithiodicarbamate.
[b]Tetramethylthiurammonosulfide.
[c]N-tert-Butyl-2-benzothiazolesulfenamide (Monsanto brand).

Despite a great deal of early interest in urethane crosslinkers as vulcanizing agents, their use in the industry has been very limited. there are several reasons for this. They are expensive and have little scorch resistance. Undoubtedly they give vulcanizates with good reversion resistance and good resistance to oxidation and heat aging. But this is in comparison to normal curing systems; thiuram systems give good reversion and aging resistance as well. From a practical viewpoint, urethane crosslinked stocks are at a distinct disadvantage because it is necessary to add a desiccant and to avoid moisture content in fillers and high humidity levels in storage.

V. METALLIC OXIDES

A. Zinc Oxide

Metallic oxides are used as vulcanizing agents for neoprene. With W types an organic accelerator is usually used as well. Three oxides are used: zinc, magnesium, and lead oxide (red lead, Pb_3O_4). The latter is used alone to improve water resistance at levels of 10–20 parts, and there is some loss in tensile strength, heat resistance, compression set, and naturally colorability. Metallic oxides modify not only the rate of cure but the scorch in neoprene compounds as well.

Zinc oxide is produced by either the American process or the French process, also known as the indirect process. In the American process zinc sulfide ore (concentrated) is oxidized and then reduced with carbon to zinc vapor. The zinc gas is drawn into a combustion zone, where it is oxidized to form pigment grade zinc oxide. In the French process zinc metal is vaporized and then oxidized in a combustion zone. French process zinc oxides generally have more nearly spherical particle shapes than American process grades and have a relatively narrow particle size distribution [8]. This type of zinc oxide is purer than American process oxide, usually assaying in excess of 99.5% zinc oxide. With a finer particle size, it is more reactive and gives fast rates of cure. Surface area of French process zinc oxide is about 7.0 m^2/g; for American process, about 4.0 m^2/g. For many purposes, however, American process material is entirely satisfactory. To facilitate incorporation and dispersion, some zinc oxides have a coating (usually about 0.5% or less by weight) of zinc propionate.

Some compounds present in zinc oxides can have effects on mixing or curing. Copper and magnesium organic salts, which could induce perishing of rubber products, do not cause problems because they are present only as inorganic salts. Lead in the pigment may react with the sulfur during cure to form dark lead sulfide and cause darkening of white or light-colored compounds. Grades averaging only 0.002% lead are considered to be lead-free. Sulfur is more prevalent in American process oxide, about 0.03%, than in French process oxide, and is considered to have a retarding effect on vulcanization.

B. Magnesium Oxide (Magnesia)

Zinc oxide is rarely used alone to cure neoprene compounds but is usually paired with magnesia, 5 parts of zinc oxide with 4 parts of magnesia being a generally suitable ratio. The magnesia used should be one of the so-called neoprene grades. These differ from common magnesias in that they are precipitated (not ground) and are lightly calcined after precipitation. They are also very active, having a high surface-to-volume ratio.

A good indication of a magnesia's usefulness for neoprene compounding is the activity number. This is the number of milligrams of iodine absorbed from solution by 1 g of magnesia. Magnesias with high activity numbers give greater processing safety, higher tensile strength, and 300% modulus, especially with neoprenes of the GN type. For example, with GN types the Mooney scorch can vary from less than 10 to more than 30 min with magnesia activity ranging from 25 to 150, respectively. In this instance the Mooney scorch is measured as a 10-point rise, using the small rotor at 250°F.

A significant loss in the efficiency of magnesium oxide in neoprene compositions can occur if it is exposed even for short periods to moisture and carbon dioxide. Magnesia is protected in transport by shipment in multiwall, moisture-resistant paper bags. It can be protected in the plant by storing small quantities in polyethylene bags until it is mixed into the compound.

REFERENCES

1. Cooper, D. L., in *Vanderbilt Rubber Handbook, 1978*, R. T. Vanderbilt Co., Norwalk, CT. p. 180.
2. Scott, A. H., *National Bureau of Standards Research Paper RP 760*.
3. *Chemical Curing of Elastomers and Cross Linking of Thermoplastics*, booklet from Lucidol Division, Pennwalt Corp. p. 8.
4. Baker, C. S. L. *Novor Vulcanizing Systems: Their Technical Development and Applicational Areas*, Rubber Manufacture and Technology Seminar, P. R. I. (Malaysian Section), Kuala Lumpur, July 21–23, 1981.
5. *Novor Application Data Sheet*, Solid Tyres, ADS-5H, Rubber Consultants, Brickendonbury, England.
6. *200 Series Urethane Crosslinkers* Bulletin 8006, Hughson Chemicals, Lord Corp., Erie, PA.
7. *Vulcanization with Novor 924*, NF. Technical Bulletin, MRPRA, Brickendonbury, England.
8. *Zinc Oxide and Rubber*, Zinc Institute Inc., New York.

8

Accelerators

I. INTRODUCTION

The reaction of rubber(s) with sulfur is slow, even at elevated temperatures. For example, a compound containing only 6 parts of sulfur and 100 parts of natural rubber would take about 4 hr to cure at 286°F. Efforts have been made to speed up the reaction for some 150 years. The first quickening of vulcanization was found when metallic oxides such as those of lead, calcium and magnesium were added to the mix. The improvement was rather small, however, and it was not until Oenslager discovered organic accelerators in 1906 that quick curing rates became available. Aniline was the first such accelerator, but its toxicity induced researchers to continue looking for alternatives. Early candidates were the xanthates (ROCSSR') and dithiocarbamates (R$_2$NCSSR). These accelerators were less toxic but have a limited market today because of their instability and low processing safety. A significant step forward was the discovery of 2-mercaptobenzothiazole (MBT) by Bedford and Sebrell [1] and by Bruni and Romani [2] in 1921. Its overall good properties spurred development, and most widely used accelerators today are of the thiazole class. Following the discovery of MBT, the disulfide mercaptobenzothiazoldisulfide (MBTS) was found to give greater safety from premature vulcanization (more scorch resistance) at higher temperatures.

Accelerator development slowed in the early 1930s, but the use of synthetic rubber and fine particle furnace black in the following decade required not only a strong accelerator but one that did not cause precure at the higher mixing temperatures involved. Such a product was introduced by Monsanto as Santocure, the brand name for *N*-cyclohexyl-2-benzothiazolesulfenamide. Competitors entered the large market that developed, and accelerators of these types were labeled delayed action accelerators because of their safe processing quality.

Since the time of Oenslager, accelerator development has concentrated on nitrogen- and sulfur-containing organic compounds. Besides accelerating the cure, these materials impart better aging to the compound. It was also found that accelerators functioned best when accompanied by metallic oxides such as lead, zinc, or magnesium oxides along with a fatty acid. Stearic acid is the one usually used, but oleic and others have been employed. Together the oxides and fatty acids are called activators and are discussed in Chapter 9.

There is a wide variety of accelerators available to the compounder. Including single compounds and blends they would easily number over 100. For ease in understanding, it is useful to classify accelerators by chemical structure. One such classification, made by the ASTM [3], is as follows:

Class 1.   Sulfenamides
Class 2.   Thiazoles
Class 3.   Guanidines
Class 4.   Dithiocarbamates
Class 5.   Thiuram disulfides
Class 6.   Thiurams other than disulfides

Possibly because of lower usage or concerns about toxicity, two classes are omitted from this list, namely aldehyde amimes and thioureas. Because of the cumbersome length of the full chemical names, acronyms are commonly used for accelerators and antioxidant (see listing in Appendix II).

Three general precautions in the use of accelerators might be noted here. Along with age resisters, they usually are the smallest in weight and volume of all the ingredients of a batch. To avoid dusting loss in mixing, the compounder can use accelerators in various forms (e.g., pellets), in elastomeric masterbatches, or attached to carrier materials. To avoid precure, accelerators usually are added at the end of the mixing cycle. This measure can also minimize loss of accelerator by volatilization. Accelerators can cause unsightly bloom on rubber vulcanizates, so their use in amounts greater than their solubility in rubber should be avoided.

Although many accelerators are single organic chemical compounds, density ranges are often quoted by suppliers. Frequently these limits are ± 0.03; for example, density of *N*-cyclohexyl-2-benzothiazolesulfenamide is given as 1.28 ± 0.03. This tolerance allows for the inclusion of various inert materials used to increase handling ease of the product. Crystal-packing density may also play a part. In compounding, the accelerator used in greatest amount is called the primary accelerator; those in smaller amounts are the secondary accelerators or kickers.

Because accelerators are so important to vulcanizate results, when a batch cures unsatisfactorily, one of the first matters checked should be whether these additives have been omitted from the batch. If a simple addition of accelerator to a small portion of stock in the correct ratio gives normal results on vulcanization, the batch usually can be saved. The stock is simply remixed with the addition of the accelerator. If abnormal results still occur when a small portion of the stock is tested, the accelerator used with that lot might be tested, perhaps simply by testing its melting point with a good reference sample. Rubber compounds can be compared on the basis of standard reference samples of certain common accelerators, obtainable from the U.S. National Institute of Standards and Technology.

The ideal accelerator (not yet developed) would be inexpensive and nontoxic, and wouldn't stain, discolor, or bloom. It would be widely compatible with other accelerators. The accelerator would be stable during all processing operations up to and including flow into a mold, but would have a fast rate of cure at normal vulcanizing temperatures to give short, economical vulcanization times.

II. THEORY OF USE

It is remarkable that for over 150 years sulfur has remained the most popular vulcanizing agent for natural rubber and general-purpose synthetics. The reactions are complex: we are dealing with elastomers having various amounts of unsaturation, as well as the action and interaction effects of (usually) metallic oxides, fatty acids, accelerators, and sulfur. Although natural rubber has a fixed amount of unsaturation, different lots can show different rates of cure dependent on the amount and kind of nitrogenous material present. Probably the most basic and far-reaching studies in the field were made by the scientists associated with the laboratories of the Malaysian Rubber Producers Research Association in England. Many of their findings were published in *Natural Rubber Science and Technology* in 1988, edited by A. D. Roberts, available from The Oxford University Press. Another comprehensive volume dealing with this field, *Vulcanizing of Elastomers*, edited by G. Alliger and I. J. Sjothun (Reinhold Publishing Corp., 1964), is a compilation of lectures on this subject at the University of Akron under the sponsorship of the Akron Rubber Group.

As a starting point, the system natural rubber and sulfur might be considered. If such a mixture is heated, probably a free radical process is involved in which the eight-atom sulfur rings break down into free radicals:

$$S_8 \rightarrow \cdot S_x \cdot$$

These free radicals unite with olefins like natural rubber:

$$\cdot S_x \cdot + 2R \rightarrow R-S_x-R$$

When sufficient polysulfidic materials are present, they serve as polysulfenyl

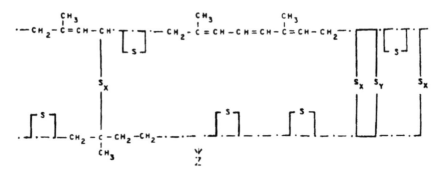

Figure 8.1 Structural features of an unaccelerated sulfur–NR vulcanizate network: Z = main chain scission. (Courtesy of L. C Bateman, *The Chemistry and Physics of Rubber-Like Substances,* Elsevier Applied Science Publishers, Ltd.)

radicals and the reaction rate accelerates. The complicated structure of the rubber–sulfur network is depicted in Figure 8.1.

Zinc oxide has several effects on rubber compositions when used as an activator at the usual 2–5 pph range. Not only does it increase the rate of reaction between sulfur and rubber, but it also changes the product distribution. Again the chemistry is complex, with the zinc oxide reacting with the accelerator as an intermediate step. Such a reaction is possible only if the zinc is in a soluble form. Solubility is achieved by including in the formulation fatty acids, such as stearic acid. These form soluble zinc salts, which then react with the accelerators. Alternatively, soluble zinc salts of fatty acids can be directly added to the compound. Another zinc salt that has been used successfully is zinc ethylhexoate.

With sulfur, zinc oxide, accelerator, and stearic acid added to the rubber, the resultant crosslinked structure is less complicated, as shown in Figure 8.2. In this system it is believed that the sulfur forms a complex with the accelerator, which in turn reacts with the rubber at the double bond or at the α-methylenic carbon to form a rubber–sulfur–accelerator complex, which then breaks down to form rubber–sulfur crosslinks. These may be mono-, di-, or polysulfidic crosslinks, and the type of crosslink depends strongly on the ratio of accelerator to sulfur. An efficient vulcanization system is one in which there is a low number of sulfur atoms per crosslink. Studies have indicated that at least qualitatively, polysulfide crosslinks occur in both NR and SBR vulcanizates. It should be clearly understood that the properties imparted to a rubber by sulfur crosslinks differ significantly from those due to the carbon-to-carbon bonds formed, for example, in peroxide crosslinking. Sulfur crosslinks, both carbon–sulfur and sulfur bonds, have lower bond energies (C-S, 66 kcal; S-S, 49 kcal) than the carbon–carbon bonds in peroxide crosslinked systems (82 kcal). For this reason the latter bonds have more thermal stability.

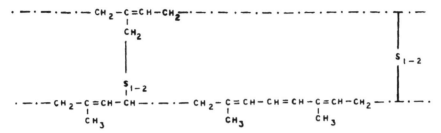

Figure 8.2 Structural features of an efficiently crosslinked MBT-accelerated sulfur vulcanizate network of NR. (Courtesy of L. C. Bateman, *The Chemistry and Physics of Rubber-Like Substances*, Elsevier Applied Science Publishers, Ltd.)

III. TYPES AND APPLICATIONS

Figure 8.3 shows the chemical structure of a representative from the accelerator classes for which a structural formula is appropriate.

Selection of a good acceleration system is one of the more difficult problems in compounding. The system chosen must have good raw stock storage stability, and its time delay before vulcanization starts must be sufficient to allow efficient mixing and processing. It must be compatible with the curing method used. A thick, rubber-covered roll, for example, may have a curing period of several hours at temperatures of 240–300°F; on the other hand, an insulated electric wire may have its rubber covering cured in a matter of seconds at 388°F. The latter temperature is that of saturated steam at 200 psi pressure used in the continuous vulcanization of rubber-insulated wiring. Once vulcanization has started, it must proceed at a satisfactory rate to an appropriate state of cure. Certain products, like steam hose, are not fully cured during the factory vulcanization. In the case of steam hose, it is subsequent use that brings the item to a full cure. Finally, the accelerator must produce a vulcanizate that gives good adhesion to fibers or metal if it is a composite article, has crosslinks that are suitable for aging or flexing, and doesn't bloom. Accelerator choice is critical, for example, in the skim coat used with steel tire cords, where strong adhesion must be obtained to the brass-coated steel wire used in radial tires.

Before reviewing individual classes of accelerators, some general observations are needed. Accelerators respond differently to different polymers. Whereas a 50-part carbon black stock of natural rubber might use 2.25 parts of sulfur and 0.5 part of a sulfenamide accelerator, a similar SBR 1500 stock might use 1.75 parts of sulfur and 0.80 part of the same accelerator. Again, rubbers like IIR and EPDM, with their low unsaturation, should not have more sulfur than can be accommodated with the available double bonds. To get fast curing, accelerator levels may be raised to a point of bloom occurs. This defect can be

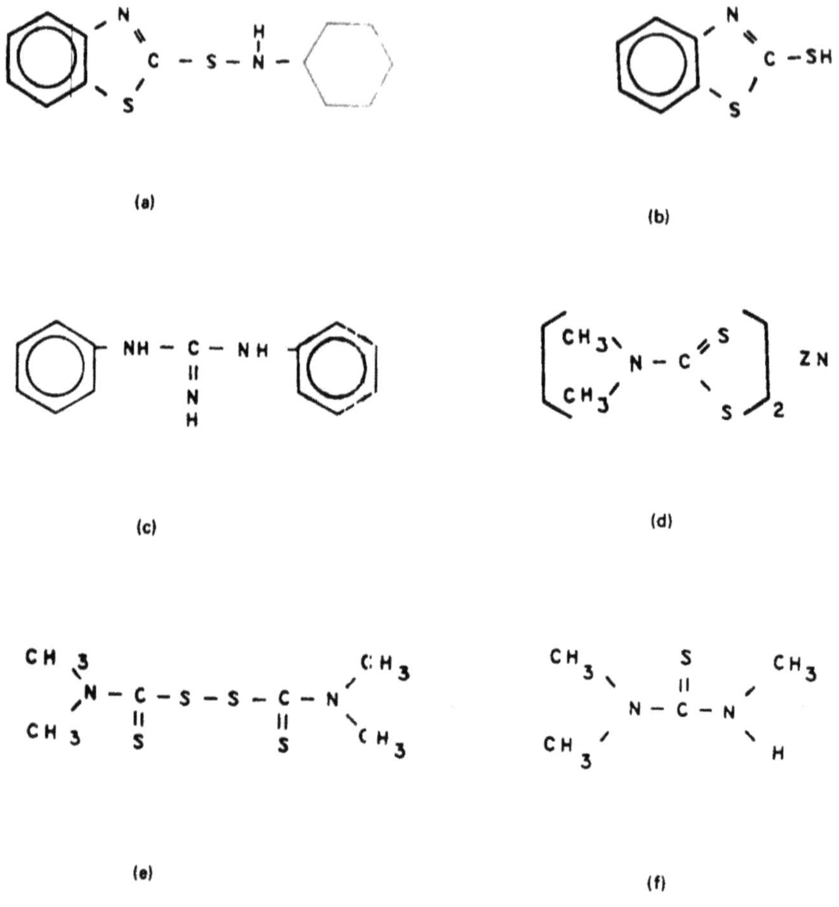

(a)

(b)

(c)

(d)

(e)

(f)

Figure 8.3 Chemical structures of accelerator classes: (a) sulfenamides, *N*-cyclo-hexyl-2-benzothiazolesulfenamide; (b) thiazoles, 2-mercaptobenzothiazole; (c) gua-nidines, diphenylguanidine; (d) di h ocarbamates, zinc dimethyldithiocarbamate; (e) thiurams, dithiuram disulfide; and (f) thioureas, trimethyl thiourea.

prevented by using small amounts of several accelerators so that the accelerator residues are soluble in the rubber. Few commonly used accelerators have sol-ubilities greater than 0.5% in natural rubber. The solubility in rubber of TMTD is approximately 0.125 Additional values are: MBTS, 0.25 and CBS, 0.5; but DPG is about 2.0%.

Specifications for accelerators ordinarily include melting point range, density, color, solubility, and frequently fineness and ash. Depending on physical form, other properties might be included such as viscosity and boiling point of liquid accelerators. 2-Benzothiazolesulfenamides are the most generally used accelerators today. They are subject to degradation during long storage, however, and in that condition their performance characteristics are impaired. A melting point determination may indicate deterioration, especially if a heating loss test is run as well. A heating loss test essentially measures moisture, and degradation of this accelerator may be caused by hydrolysis. Some accelerators have higher melting points than usually occur in mixing. To achieve even dispersion, the powder form should be used. Some evidence of dispersability is obtained by wet sieving the accelerator—obviously if this is difficult, dispersibility is questionable.

A. Sulfenamides

The latest significant development in accelerators occurred in the 1930s when Zaucker and Bogemann discovered that sulfenamides were delayed-action accelerators of vulcanization [4]. Then in 1937 Harman discovered *N*-cyclohexylbenzothiazole-2-sulfenamide [5]. The delayed action this type of accelerator possessed became particularly in demand with the introduction of scorchy oil furnace blacks in the 1940s. Since then developments have been directed to developing still greater delayed action.

Sulfenamides are made in two ways: by the reaction of 2-mercaptobenzothiazole with an *N*-chloramine and by oxidation of the appropriate amine salt of 2-mercaptobenzothiazole. Most suppliers promote one or more sulfenamide-type accelerators. The original *N*-cyclohexylbenzothiazole-2-sulfenamide, marketed by Monsanto as Santocure, is sold as Vulkacit DZ by Bayer, as Vulcafor HBS by Imperial Chemical Industries, and as Durax by Vanderbilt.

The various sulfenamides reflect in their scorch times and cure rates differences in the amines used in their manufacture. If the amine is highly basic, the scorch time is lower, the cure rate faster. In terms of processing safety, *N,N*-diisopropyl-2-benzothiazolesulfenamide (e.g., American Cyanamid's DIBS) is probably the highest, with *N*-cyclohexylbenzothiazole-2-sulfenamide the lowest. The trend is to use the modifications of the original sulfenamide accelerators that give greater processing safety. A popular accelerator is now *N*-*tert*-butyl-2-benzothiazolesulfenamide (TBBS). Besides giving a longer scorch delay time, this product gives higher modulus than CBS, so dosage can be reduced 10%.

In a 50-part HAF black natural rubber stock, 0.5 part of TBBS with 2.5 parts sulfur gives a Mooney scorch time of approximately 13 min. This result is the time required for a 10-point rise in the Mooney viscosity at 275°F. In an equivalent SBR stock with a curative system of 1.75 parts sulfur and 1.25 parts TBBS, the scorch time under the same conditions is about 25 min. Very frequently the sulfenamides are used with secondary accelerators like TMTD. Starting curing formulations in an NR stock might be 0.70 part sulfenamide,

0.10 TMTD, and 2.25 sulfur. In an SBR stock the proportions would be 0.90 sulfenamide, 0.10 TMTD, and 1.90 sulfur.

B. Thiazoles

By far the most popular accelerators are the thiazoles and derivatives of the thiazoles, the sulfenamides. Probably over 70% of the accelerators used are of these types. Their good qualities are the effective acceleration they provide at medium and high temperatures and their wide range of curing rates and scorching characteristics. Quite frequently they are paired with a secondary accelerator to derive maximum levels of processing safety and vulcanization rate.

Raw materials for this class of accelerator are aniline, carbon disulfide, and sulfur. The chief thiazole accelerators are 2-mercaptobenzothiazole, mercaptobenzolhiazole disulfide, and the zinc salt of 2-mercaptobenzothiazole. Two of the oldest names in accelerators, going back to the 1930s, are Captax and Altax, Vanderbilt's brand names for 2-mercaptobenzothiazole (MBT) and mercaptobenzolhiazole disulfide (MBTS). The zinc salt is supplied by several manufacturers, including Monsanto, which makes Bantex. The zinc salt is rarely used in dry compounding but is used more in latex foams and dipped goods. Both MBT and MBTS can easily cause scorching problems when used alone in furnace black natural rubber stocks. MBTS is somewhat safer to process than MBT. It is common practice to use a kicker with MBTS and MBT. MBTS is often used with DPG to give a safe, flexible system in many mechanical goods stocks. A starting curative system for an SBR stock might be 1.5 MBTS, 0.15 TMTD, and 2.80 sulfur.

C. Guanidines

The two main guanidine accelerators are diphenylguanidine (DPG) and di-*o*-tolylguanidine (DOTG). The guanidines are now rarely used as primary accelerators because of their slow curing. They can be used as primary accelerators in natural rubber roll-covering stocks, where long cure times are used at comparatively low temperatures to assure the most uniform state of cure throughout the thick covering. Their main use s as a secondary accelerator in thiazole- or sulfenamide-accelerated natural rubber or SBR stocks. A common ratio is 0.25 part DPG to 1 part of thiazole or sulfenamide accelerator. Although both are white powders, guanidines can stain to some extent so they are not used in the best white or light-colored stocks

D. Dithiocarbamates

Dithiocarbamates, the metal salts and amine salts of dithiocarbamic acids, are called ultra-accelerators because of their quick curing characteristics. Popular members of this class are the zinc methyl and ethyl dithiocarbamates. Salts of bismuth, cadium, copper, lead, selenium, and tellurium also are used. An accelerator like selenium dimethyldithiocarbamate is useful in low sulfur or sulfurless cures (discussed later).

Dithiocarbamates are so powerful that they are rarely used alone except in such specialties as spread goods (fabric covered with a rubber coating: e.g., hospital sheeting) cured in air at or slightly above room temperature. Usually they are paired with thiazole or sulfenamide accelerators to adjust the cure rate of a stock. A typical curing system with natural rubber might be 0.5 part of zinc dimethyldithiocarbamate (ZMDC), 0.75 part of a thiazole accelerator, and 2.0 parts of sulfur. With SBR stocks the system might be 0.6 ZMDC, 0.75 thiazole, and 1.8 sulfur. Usually nonstaining, the thiocarbamates are versatile accelerators and are used in IIR and EPDM as well as NR and SBR. Typical commercial brands are Vanderbilt's Ethyl Tellurac (tellurium diethyl dithiocarbamate, TDEDC), Bayer's Vulkacit LDB (zinc-*N*-dibutyldithiocarbamate, ZBDC), and Monsanto's Methasan (ZMDC).

E. Thiuram Mono- and Disulfides

Like the thiocarbamates, thiuramsulfides are ultra-accelerators. They are made from secondary amines and carbon disulfide. The most commonly used members of this class are probably tetramethylthiuramdisulfide (TMTD), tetraethylthiuramdisulfide (TETD), and tetramethylthiurammonosulfide (TMTM). The strongest of these three is the tetramethyldisulfide, followed by the tetraethyl and then the monosulfide. TMTD is too scorchy to be used alone with natural rubber. TETD is about the same, but the monosulfide can be used with care. If TETD is substituted for TMTD, about 10% more accelerator should be used. Like the dithiocarbamates, the thiuramsulfides can be used in light-colored stocks without staining.

The thiuramsulfides are especially useful in curing systems containing low or no elemental sulfur, but the monosulfide is not strong enough to cure satisfactorily without added sulfur. The most common use of the thiuramsulfides is as a secondary accelerators to thiazole- or sulfenamide-accelerated compounds of NR, SBR, IIR, and EPDM. For example, an EPDM stock might have a cure system of 1.5 parts sulfur, 1.25 parts thiazole, and 1.5 parts TMTD. Typical commercial accelerators of this class are Monsanto's Monothiurad (TMTM), Imperial Chemical Industries Vulcafor TMTD (TMTD), and Vanderbilt's Ethyl Tuads (TETD).

F. Thioureas

The thiourea accelerators are almost exclusively used as organic accelerators in neoprene stocks. Among producers of these accelerators are DuPont with NA 22F (ethylene thiourea, ETU) and Vanderbilt with Thiate H (*N*,*N'*-

diethylthiourea). These accelerators are particularly useful in injection-molded stocks, molded sponge, and liquid curing medium (LCM) extrusion stocks. Customary dosage is 0.5–0.75 on the rubber, usually neoprene W.

G. Aldehyde Amines

Aldehyde amines are reaction products of various aldehydes and primary amines. They vary from brown resins to amber liquids and are so complex that structural formulas would have little meaning. A common combination of starting materials is butyraldehyde and aniline, which gives such commercial products as DuPont's Accelerator 808 and Imperial Chemical Industries's Vulcafor BA. Another accelerator in this class is Heptene Base, made by Naugatuck Chemical from heptaldehyde and aniline.

Aldehyde amine accelerators now have limited use. The curing rate they impart depends on the choice and ratio of aldehyde and amine. These accelerators can be used to give hard, nonbrittle stocks like ebonite. For a natural rubber ebonite, perhaps 2.5 parts of this accelerator might be used with 30–50 parts of sulfur. A more common use is as a secondary accelerator at 0.20–0.30 part with thiazole or sulfenamide accelerators.

IV. SPECIAL APPLICATIONS

Over the years three special types of organic accelerator cure system have been developed, primarily for natural rubber but of value in SBR as well. They are efficient cure systems, semiefficient cure systems, and soluble cure systems. Much earlier, when very little was known about vulcanizate structures, the first was called sulfurless curing, the second low sulfur curing. Soluble cure systems are a relatively recent development. In general, efficient cure systems use no elemental sulfur; the sulfur for vulcanizing comes from sulfur donors such as thiuramdisulfide accelerators. Vanderbilt's Sulfads, essentially dipentamethyl-enethiuramhexasulfide, has 35% available sulfur, the more common TMTD accelerator about 13.3%. In semiefficient cure systems the amount of sulfur used is generally less than 1 part.

The advantage of efficient and semiefficient vulcanization systems in both NR and SBR compounding is the increased resistance to aging and reversion (deterioration caused by overcuring). Soluble curing systems, a variant of efficient vulcanization systems, are designed to ensure high reproducibility from mix to mix and to improve resistance to creep or stress relaxation. The performance of a sulfur vulcanizate is the result not only of the system used but also the solubility of the components in the unvulcanized stock and the solubilities of the reaction residues. If the amounts of sulfur, accelerators, or fatty acids used are above their solubility limits in rubber, these materials can crystallize out within the rubber. Excessive crystallization can result in inhomogeneity in crosslinking and variations in vulcanizate behavior. Such disadvantages can be minimized or eliminated by using ingredients that are soluble in the rubber in the concentrations used. This area has been explored by Elliot, Smith, et al. [6–8].

Many studies have documented both the advantage (increased age resistance), and the disadvantage (impaired fatigue resistance) as measured by flex cracking of efficient and semiefficient cure systems. In one study [9] a 50-part HAF black loaded stock showed a drop in tensile strength on aging (48 hr at 212°F) of 33% when a conventional curing system was used. When an efficient curing system was used, the drop was only 23%. The drops in elongation under the same aging conditions were 40 and 16%, respectively. In another study [10] in natural rubber the efficient vulcanization system stock retained 91% of its original tensile strength after aging 12 days at 85°C, a semiefficient 79%, but a conventionally cured stock only 41%. However, when fatigue was measured by extending the stock to 100% extension and return, the conventional stock lasted for 74 kilocycles, the semiefficient 27 kilocycles, and the efficient 29 kilocycles.

It is evident that there are tradeoffs in the use of efficient vulcanizing systems. Besides the technical tradeoff of improved aging but inferior fatigue resistance, there are cost considerations. Specifics are hard to come by, but a 10% increase might be expected.

The reasons for the effects on aging and flex-cracking resistance of these curative systems are not completely understood. The nature of the crosslinks formed during vulcanization with the various systems and the changes that take place on aging may well be important factors. The nature of the crosslinks and their distribution can be determined by using chemical problems like lithium aluminum hydride. Conventionally cured natural rubber compound will have no monosulfidic crosslinks but approximately 30% disulfide and 70% polysulfidic crosslinks. On postvulcanization aging after vulcanizing changes occur, the amount of sulfur bound to the network as cyclic disulfides increases and the polysulfidic crosslinks change to monosulfidic crosslinks. More chain scission occurs, which causes tensile strength and elongation to deteriorate, and oxygen attack begins. In efficient vulcanizing systems the main crosslinks are mono- and disulfidic. Crosslinks of these types do impart good thermal stability. However, it is thought that they lack the flexibility to rearrange themselves to cope with the localized stresses associated with a flaw, hence the impairment in flex life. Perhaps polysulfidic crosslinks have this ability.

V. ACCELERATORS AND TOXICITY

Recently there has been concern about the toxicity of nitrosoamines ($-R_2$ $NN=O$) in rubber products. Such compounds can be brought into the mix by certain accelerators, or they can be formed during vulcanization or develop in stored rubber goods. The amounts are very small, but modern analytical techniques can detect amazingly small qu antities. One of the severest restrictions on *N*-nitrosoamines in the atmosphere s that mandated in Germany of less than 2 μg per cubic meter of air. Nitrosoamine s cause cancer in animals during laboratory testing and are considered to be possible human carcinogens. A fundamental investigation of the problem is that of Seeberger and Raabe [11].

Nitrosoamines develop from accelerators when the latter are made from secondary amines Two popular classes of accelerators made in this way are the more scorch-resistant sulfenamides such as DIBS (*N,N*-diisopropyl-2-benzothiazolesulfenamide) and the ultra-accelerators, thiuram sulfides, such as TMTD (tetramethylthiuramdisulfide).

Because of the suspected carcinogenicity mentioned above, and ensuing government regulations, accelerator manufacturers such as Monsanto and Akzo have developed substitutes.

Monsanto has worked on modifying its Santocure NS brand (*N-tert*-butyl-2-benzothiazolesulfenamide). As a primary amine derived sulfenamide, it does not form stable *N*-nitrosoamines. From this work they have developed Santocure TBSI, which is an imide and incorporates another mole of mercaptobenzothiazole as shown.

Chemical Name   N-t-butyl-2-benzothiazolesulfenimide

Generic Abbreviation   TBSI

This product has been tested in NR and SBR compounds, and the results for processing and curing characteristics fall within the population from existing secondary amine derived sulfenamides. In the natural rubber compounds Santocure TBSI showed a modest improvement in reversion resistance compared to the sulfenamides. Incorporating a mole of MBT to form a sulfenimide greatly increases the sensitivity to hydrolysis, and the storage stability of his new accelerator is excellent. Details are provided by Luecken and Lederer [12].

Akzo has developed an accelerator to replace accelerators of the thiuram class (e.g., TMTD). Having found from published literature that dibenzulnitrosoamine is not carcinogenic, the company introduced Perkacit TBzTD (tetrabenzylthiuramdisulfide) as an alternate accelerator. The vulcanizate properties achieved with this accelerator were much the same as those of TMTD in NR, SBR, NBR, and EPDM. The replacement, though, entails certain compound changes. Long scorch times and slow curing rates are attributed mainly to the amount of available sulfur in TBzTD, which is about half that of TMTD, since the molecular weight for TBzTD is about twice that of TMTD (544 vs. 240). In some cases the deficiencies noted can be corrected by increasing the free sulfur in the compound. Akzo comments that in general addition of an extra 10% sulfur calculated on the intake of TBzTD in an efficient vulcanization system is sufficient to achieve a crosslinking comparable to that of TMTD. Strangely, increased sulfur is not required in SBR compounds. If increasing the sulfur does not bring about an equivalent curing rate, addition of guanidine accelerators might help. Although TBzTD contains a low amount of NDBzA (dibenzyl nitrosoamine), concentration increases minimally during vulcanization and is nonvolatile during storage of rubber parts at normal temperature. Further data on the use of TBzTD is given in two Akzo brochures [13,14].

Although this section relates only to accelerators, nitrosoamines are formed from the use of other ingredients in rubber compounding. Perhaps down the road rubber goods plants will require a consultant toxicologist to anticipate and satisfy regulations.

Activators and Retarders

I. ACTIVATORS

Although almost all rubber compounds have activators, retarders are used spar-
ingly. Activators have been described as substances that increase the effects of
accelerators. Retarders are ingredients added to a stock to prevent its premature
vulcanization under processing conditions. By far the most popular activator
system is zinc oxide and stearic acid. The mechanism by which this pair speeds
up the cure has not been exhaustively studied, and results to date indicate its
complexity. It would seem that zinc oxide reacts with the stearic acid to form
zinc stearate, which is soluble in the rubber (the latter might be considered to be
an organic solvent) and in this form facilitates the crosslinking process. It is
essential to have the zinc ions in soluble form.

   The presence or absence of the common activators in a simple natural rubber
stock has been examined by Paris [1]. The results appear in Figure 9.1. This
graph shows the changes in torque when a stock with the variations listed in
Table 9.1 is cured in a biconical disk curemeter (the Monsanto rheometer). The
very slow curing of rubber with sulfur alone is not improved very much by the
addition of zinc oxide and stearic acid. When sulfur and a sulfenamide accelerator like N-tert-butyl-2-benzothiazolesulfenamide (TBBS) are added alone to the
rubber, a small but significant state of cure is developed quickly. The rate and
state of cure of that mix is only marginally improved by the addition of stearic
acid. If the stearic acid is omitted but zinc oxide is present, there is a much faster

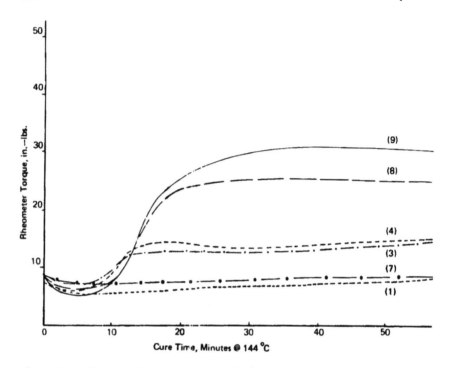

Figure 9.1   Effect of activators on cure rate; for key to numbered curves, see Table 8.1.

cure rate. It is known that natural rubber itself contains some fatty acids. Finally, if the cure system includes sulfur, zinc oxide, stearic acid, and the sulfenamide accelerator, the fastest rate of cure is developed. Certain vulcanization accelerator systems, however, do not require zinc oxide; these include systems where the accelerators are aldehyde amines or DPG.

A common feature of activators is that they are alkaline and besides zinc oxide include litharge, magnesia, and amines. Even furnace blacks have an activating effect. Litharge is an activator for *m*-dinitrobenzene cures of SBR; red lead (Pb$_3$O$_4$) is the preferred activator for GMF cures of butyl stocks. A disadvantage of using lead oxides in sulfur-vulcanized stocks is the discoloration produced by the formation of black lead sulfide.

Historically it was general practice to add 5 parts of zinc oxide when ZnO was used as an activator. More recently the tendency has been to use 2–4 parts. Studies have indicated there is little if any gain in activation function beyond 4 parts, and 3 is adequate for most purposes. The reduction is worthwhile because zinc oxide is a relatively expensive pigment and has a high density. There is a more recent reason for reducing zinc oxide dosage. The U.S. Environmental Protection Agency (EPA) has determined that zinc oxide is a toxic chemical;

Table 9.1 Key to Curves in Figure 9.1

Stock	Cure system
1.	Sulfur (2.5 pph)
7.	Sulfur + stearic acid (2 pph) + zinc oxide (5 pph)
3.	Sulfur + TBBS (0.6 pph)
4.	Sulfur + TBBS + stearic acid
8.	Sulfur + TBBS + zinc oxide
9.	Sulfur + TBBS + stearic acid + zinc oxide

Note: TBBS is *N-tert*-butyl-2-benzothiazolesulfenamide.

accordingly, the less used, the less difficulty in disposing of waste rubber compounds. The properties and uses of zinc oxide are described more fully in the Chapter on nonblack fillers.

The other half of the activator couple is the fatty acid used, although the two can be combined by adding the zinc salt of an appropriate fatty acid. This, for example, might be zinc laurate (e.g., which has 0.2 parts of zinc oxide in rubber-soluble form). Usually the fatty acids employed have 12–18 carbon atoms. Some curing systems do not require fatty acids such as no-elemental-sulfur curing groups using tetramethylthiuramdisulfide. Again they are not used with peroxide-cured compounds.

Polymers themselves differ in their fatty acid contents. Natural rubber has fatty acids which are probably equivalent on the average to 2 parts of stearic acid. This, however, does vary so it is customary to add additional acid. If emulsion process SBR is used it may well contain fatty acids from the coagulation process, but it may need more. (Fatty acid soaps are used as emulsifiers in many SBR types.) The other rubbers reviewed here—CR, IIR, T, NBR, BR, and EPDM— do not contain fatty acids. The curing system for neoprene is different and does not require fatty acids.

The amount of stearic acid used will vary from 1 to 3 parts on the rubber. For the relatively lower unsaturation of EPDM, 1–2 parts is sufficient. Besides its role in activation, stearic acid acts as a lubricant in rubber mixes, reducing the viscosity. Rubber grade stearic acid is not a chemical compound, but rather a mixture of fatty acids, typically containing about 60% stearic, 21% palmitic, and smaller amounts of myristic and oleic acids. For critical compounding, the amount of unsaturation in the stearic acid is important. Stearic acid high in unsaturation may react with the sulfur in the compound and thus reduce the vulcanization of the polymer.

For this reason the ASTM has two classes for stearic acid. Class 1 stearic acids, used where unsaturation would not be a problem, and has three grades, depending on iodine value: low allows iodine values up to 8, medium up to 15, and high to 39 ± 5. Class 2 requires iodine values less than 1.0 for two grades: in grade one the ratio of palmitic to stearic is 50:40, and in grade two it is 30:65. The iodine value or iodine number of an oil or fat is a measure of its unsaturation. In this test a solution of iodine and iodine chloride is usually used, and the amount added to the olefinic material in carefully specified conditions is determined. Typical unsaturated fatty acids are oleic and linoleic.

Other materials used to provide fatty acids include coconut, palm, and cottonseed oils. Chief fatty acids in these are as follows: coconut oil: 48% lauric, 17% myristic; palm oil: 43% oleic and 42% palmitic; cottonseed oil: 45% linoleic, 29% oleic, and 21% palmitic. Vegetable oil prices vary widely, and can be a reason for their selection over stearic acid.

II. RETARDERS

Whereas the value of activators is quite evident, that for retarders is not so clear. Their use does increase safety in processing operations, and prematurely vulcanized stock is difficult if not impossible to salvage. On the other hand, effective retarders slow down the cure rate as well, so there are tradeoffs. Retarders currently are of two kinds—acidic retarders and cyclohexylthiophthalimide (CTP), the latter marketed by Monsanto as PVI. The acronym PVI stands for prevulcanization inhibitor. A third type, *N*-nitrosodiphenylamine, is being discarded for health safety reasons. The acidic retarders include salicylic acids, benzoic acid, maleic acid, and phthalic anhydride.

Considering the acidic retarders first, like so many other aspects of vulcanization, the way they function is not clear. It is possible that the organic acids may react with basic impurities in the formulation which would otherwise accelerate the onset of cure. They are rarely used at dosages over 2 parts and, as mentioned earlier, definitely decrease the cure rate.

Over the years, many methods for measuring cure rate have been proposed. Perhaps the most popular are those derived from using an oscillating disk curemeter. A prime example of this is the Monsanto rheometer. The biconical disk is oscillated at a small rotational amplitude in a cavity filled with the rubber under test. The force required to oscillate the disk to maximum amplitude is continuously recorded as a function of time. Ordinarily the torque first drops as the rubber becomes more plastic with heat, then rises as crosslinking occurs, and finally reaches a maximum value. The test is described in ASTM D-2084-91 [2]. The cure rate can be defined as the ratio:

$$\frac{100}{T_x - t_{s_1 \text{ or } 2}}$$

where T_x is the time in minutes required for a stock to reach $x\%$ of the maximum torque and $t_{s_1 \text{ or } 2}$ is the time in minutes for the torque to rise 1 or 2 lb/in. above the minimum torque. The usual values are 90% for x and t_{s_2}; t_{s_1} is used when the oscillating disk rotates through a 1° arc and t_{s_2} when the arc is 3°.

Cure retardation by retarders is easily determined using such rheometers, and the cure index of the stock with and without retarders can be found, as well.

Obviously, slower cure rates mean longer cure times and consequently lower productivity out of the molds and higher costs. With 1.5 parts of phthalic anhydride in a natural rubber black formulation accelerated with MBTS/DPG, the cure index values might drop from about 12 min^{-1} to about 8 min^{-1}.

The choice of a retarder is determined by considering the rubber used, the accelerator system, the processing temperatures, and the amount of processing involved. Acidic retarders, usually used with thiazole-accelerated stocks, are not effective in sulfenamide-accelerated stocks.

In essence the compounder should try to develop a curative system that renders retarders unnecessary. In doing this a thiazole–DPG coupling might be replaced with one of the sulfenamide accelerators. If this doesn't work out, the possibility of using CTP should be considered.

Among the advantages of CTP is its effectiveness with a wide range of polymers, accelerators, and other compounding ingredients. It neither affects vulcanizate properties nor causes staining or porosity. Although most effective with sulfenamide-accelerated stocks, it is used with both MBT and MBTS. With its use that valuable balance of processing safety, cure rate, and vulcanizate qualities is more easily obtained. There is a linearity about its response that allows one to determine the dosage necessary to give a certain scorch resistance. In most applications 0.1–0.3 part is used. Certain reservations are necessary. CTP is not effective in either resin or metal oxide curing systems, and in butyl stocks the response lags and dosage must be adjusted upward from the range just given—perhaps up to 1 part.

REFERENCES

1. Paris, W. W., Accelerators, ACS Rubber Division Compounding Symposium, Fall 1982, Chicago.
2. *1991 Annual Book of ASTM Standards*, Vol 09.01, p. 391.

10

Reinforcers: Carbon Blacks

I. INTRODUCTION

With the exception of sulfur, no material has increased the usefulness of rubbers as much as carbon black. With its wide variety of production processes, however, a precise definition is difficult. A working definition could be that carbon black is an amorphous carbon of quasi-graphitic structure. Most of the carbon black produced is used in rubber products, although an important secondary application is in printing inks. In the plant the black is first produced as a fluffy powder, in which form it is readily usable in inks and protective coatings. However most is further processed into pellets for easier handling by rubber goods manufacturers.

The primary purpose for using carbon black with rubber is to reinforce it. By reinforcing is meant the enhancement of tensile strength, modulus, abrasion, and tear resistance obtained by adding the material. The excellent wear resistance of rubber with carbon black was first noticed in England in about 1912. Until then, the normal filler used in tire treads was zinc oxide, and tire mileage was a matter of a thousand or so miles. Substitution of carbon black for zinc oxide increased the mileage many-fold. Improvements in carbon black and in the way it is used, along with improvements in tire design and manufacturing, led to present-day passenger tires good for 40,000–80,000 miles. Specific tire life is impossible to predict, since road conditions, vehicle weights, driver characteristics, and other factors have significant effects on tire durability.

At present probably 30 ± 5 lb of carbon black is used for every 100 lb of rubber. The development of oil furnace blacks freed carbon black producers from being tied to nearby natural gas fields. As a result, for reasons of self-sufficiency, national pride, and economics, production spread from the United States to all quarters of the globe. Present-day industry research appears to be concerned not with developing new production processes but with studying the nature of black itself, determining what useful new grades might be produced on existing equipment, and investigating such measures as energy conservation.

II.   PROPERTIES

A.   Physical Properties

The blackest pigments available for printing inks and protective coatings are carbon blacks. Carbon blacks in their fluffy form have the smallest particle size of any industrial commodity. For classification purposes the ASTM considers the particle size range of carbon blacks used in rubber products to be from 1 to 500 nanometers (nm). A namometer is only 10^{-9} m. Most rubber grade carbon blacks fall in the 20–26 nm range.

Carbon blacks are hygroscopic and a black of high surface area, such as Super Abrasion Furnaces [the former industry term is still widely used] could easily pick up 2 5% moisture if exposed to high relative humidity. A common value used for the density of carbon black in rubber is 1.80. This may not be accurate enough for some highly specified products.

One black producer [1] has listed specific gravities for their products ranging from 1.761 to 1.803. What matters is the density of the black when it is mixed in rubber. One way to determine this is to carefully mix and cure two batches, one containing half the amount of black that the other has. It is assumed that dust and mixing losses for the two batches are equal, so the difference in density is caused by the difference in weight of black, and from this the density of the black in rubber can be calculated.

In the pelleted form the pour density can vary from 16.0 to 40.0 lb/ft^3. Besides being a quality-checking test, these values are useful in the design of bulk storage tanks.

Physical properties of carbon blacks affect both rubber processing and vulcanizate properties significantly. Because of the fineness of this material and the variety of conditions under which it is produced, precise descriptions of its morphology are hard to make. Descriptions have been developed largely through the use of the electron microscope. It is interesting that a carbon black, Shawinigan acetylene black, was one of the first materials studied at the University of Toronto, where the first electron microscope on this continent was developed. Figure 10.1 is a transmission electron micrograph of a medium thermal (MT) black, Thermax. Figure 10.2 is another electron micrograph of a high structure

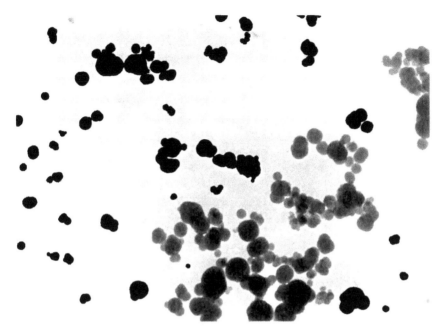

Figure 10.1 Medium thermal (MT) black at 20,000 × magnification. (Courtesy of Cancarb Ltd.)

reinforcing black, Cabot's Vulcan 4H. When the differences in magnification are noted, the fineness of the reinforcing furnace black compared to the medium thermal is readily apparent. Other differences are covered later.

Properties that have the most influence on rubber processing and vulcanizates are particle size, surface area and character, and structure. When formed in the furnace, carbon black particles are spherical. This spherical configuration largely carries through for thermal blacks to the end product. Other blacks, especially the popular furnace blacks, tend to form chains or networks of these particles fused together. Electron micrographs of thermal blacks indicate that the particle size distribution is skewed to the larger size, and this is generally true for other grades.

Special techniques have been developed so that the size of the particles, or the nodes as they are sometimes called, can be estimated from observations on the aggregate. Within an aggregate of carbon black, the carbon atoms in the node are in imperfect graphite layers generally parallel to the surface. The layer planes can extend from one particle or node to another. Although individual separate particles are rarely found in blacks other than thermal blacks, the surface average diameter of the particles can be related to the properties of the carbon black

Figure 10.2 High structure reinforcing black Vulcan 4H, at 100,000 × magnification. (Courtesy of Cabot Corp.)

rubber compound. Thermal blacks are not porous, so the surface area is inversely proportional to the diameter. Compared at equal loadings to other rubber blacks (all finer in particle size), thermal blacks give more plastic mixed stocks, as well as vulcanizates with lower hardness with little reinforcement but high rebound.

Most rubber grade blacks are oil furnace blacks. Here the individual particles are smaller, and there are few isolated particles visible in electron micrographs. It is believed that in the reactor, the fine oil droplets lose hydrogen, become more and more viscous, collide, and stick to one another. With the smaller contacting particles, surface area of the black per gram increases despite the loss of surface at the contact bond.

Probably the most important property of a carbon black is the surface area that is accessible for reaction with rubber molecules per unit of weight, usually in square meters per gram. The term "accessible" is used because some blacks have pores whose area is included in surface area measurements but cannot be reached by rubber molecules. Surface area of rubber grade carbon blacks determined by electron microscope techniques can vary from 9 to 143 m^2/g [2]. The chemistry of the carbon black surface is complex, and much more research must be done before its influence on mixing, curing, and vulcanizate properties can be well understood. Some features are known, channel blacks, now little used for compounding, had volatile contents of 3–4%, which caused the black to be acidic and slower curing compared to furnace blacks with very little volatile content.

The third significant physical property is structure. The carbon black aggregate does not resemble a segmented ceramic snake. The nodes or particles can be tightly clustered together like a bunch of grapes, or the same number of nodes can be arranged in more open fashion, giving greater bulk, in a manifestation called high structure. Structure can be modified by varying furnace conditions and feedstocks. Among other changes, high structure blacks give increased viscosity to uncured stocks and increase the smoothness of extrusions. Increasing structure in vulcanizates increases hardness, abrasion resistance, and electrical conductivity, but decreases cut growth and flex resistance.

Determination of particle size, with its dependence on electron microscopy, requires facilities beyond the reach of most compounders. Surface area determination is probably more important, as theories suggest that reinforcement varies directly with the surface area available for rubber–black interaction. The ASTM iodine number test [3] is frequently used for surface area determination. A relatively simple test, it measures the amount of iodine adsorbed by the black from a potassium iodide solution. The iodine number test is not suitable for blacks with a high volatile content or high amounts of extractable matter. (Fine Thermal black, no longer marketed, had about 3% extractables.) For such blacks, adsorption of cetyltrimethylammonium bromide (CTAB) from an aqueous solution of the black is preferred. This method is more difficult to run than the iodine number test, and it too is described in an ASTM test [4].

Structure is commonly measured by slowly adding a wetting liquid like linseed oil or dibutyl phthalate to the black under carefully controlled mixing conditions until the black forms a smooth, coherent ball. This time-honored test has been refined and developed by the ASTM into an automated procedure based on the so-called dibutyl phthalate absorption number of compressed sample [5]. The end point is taken when there is a sharp rise in viscosity. The higher the amount of dibutyl phthalate absorbed, the greater the structure. Pelletizing changes the absorptive nature of a black. Accordingly, the black is compressed before being used in this test to more closely match the condition of the black when it is dispersing in the mixer.

Carbon blacks are valued for the blackness they impart to plastics, printing inks, and protective coatings. The light absorption coefficient is commonly measured by the tinting strength test. In this test the carbon black sample is mixed with zinc oxide and soybean oil epoxide, which serves as the liquid medium. When the paste has been spread satisfactorily, it is measured for reflectance by a photoelectric reflection meter. The reflectance is then compared with that of a similar paste made with a reference black and the relative tinting strength determined. Although tint strength is highly dependent on particle size, there is also dependence on structure and aggregate size distribution. The range of blackness that occurs in carbon black spectrum was noted by Lyon and Burgess [6], who pointed out that over 250 shades of black distinguishable by the naked eye can be produced in mineral oil pastes by changing from the coarsest to the finest carbon black.

Thanks to more highly controlled processes and in some cases improved reactor design, several important blacks were developed roughly in the period 1972–1977. At least one supplier has referred to them as new technology blacks. Earlier, blacks used for tire treads, a most important market, had surface areas from about 80 to 140 m^2/g, determined by iodine number. Because of processing problems (good dispersion was difficult), it was impossible to take full advantage of the greater wear possibilities of the high surface area blacks—say 125–140 m^2/g. Tire makers were demanding higher tread wear mileage, yet at the same surface areas by iodine numbers that they commonly used.

The new blacks were an answer to such requests. It is perhaps an oversimplification, but the major change was that that the new blacks, while maintaining iodine surface areas at the same levels, had significantly higher external surface area by N_2 absorption methods and gave higher tread wear ($\approx$ 5–10%). Such a black might have an iodine surface area of 90 m^2/g but a nitrogen surface absorption surface area of 120. Other properties that were changed were structure (higher structure as measured by the dibutyl phthalate absorption test) and tinting strength (also higher).

These new blacks have a narrower distribution of aggregate size, and their surfaces are more reactive, as shown by higher moisture absorption. Modified carbon blacks are continually being developed, largely as a result of constant pressure from auto makers. Attempts to conserve oil by increasing mileage per gallon have led to a demand for tires with lower rolling resistance. At the same time, high tread wear resistance is expected, as well as good handling and cornering, and superior traction on wet and dry roads. Typical of research in these areas is that by Columbian Carbon, which has developed the experimental black CD 2005. A compressed sample of this very high structure black had a dibutyl phthalate absorption of 132 cm^3/100 g and a surface area of 121 m^2/g, as determined by electron microscopy. The black showed unique compounding flexibility. At the same modulus as other black compounds, loading can be lower, with an accompanying reduction in rolling resistance. Additional details are given by Swor et al. [7].

B. Chemical Properties

It can be assumed that carbon black used in rubber assays 95% carbon at a minimum. Volatile content, usually moisture, can be around 2.5% for high surface area blacks. Ash content increases with surface area and may reach 0.75–1.0%.

Although undoubtedly reinforcement is influenced by the chemical reactions between black and rubber, chemical properties of blacks are not of direct interest to the compounder. Carbon blacks are relatively inert but can be oxidized by concentrated nitric acid. Surface gaseous groups usually render a water slurry of a channel black acidic; oil furnace blacks with little volatile content usually have a pH of 7.1–9.0. Ash content is low, and a significant portion of it is residual salts from the cooling and pelletizing water.

Occasionally the question of the sulfur content of carbon black comes up—specifically, whether the sulfur should be considered to be a part of the vulcanization system. Over several decades, the sulfur content of refinery streams available as carbon black feedstocks has risen appreciably. As a result, some carbon blacks may have as much as 1.5% sulfur. However, studies have shown it to be trapped, as it were, in the crystal lattice of the black, such that its vulcanizing effect is negligible.

C. Health and Safety

Carbon black is one of the safest materials used in a rubber goods factory. Unlike some other fine particle materials, such as grain dust, it does not explode spontaneously. It is ignited with difficulty and supports combustion very slowly. If a fire starts in bagged carbon black storage, simply separating the bags in the pile will bring the process to a halt, and damaged bags can be wetted down and discarded.

It is true that with prolonged extraction by certain solvents a very small amount of residual tarry material can be extracted which contains polyaromatic hydrocarbons. Such materials have been suspected of being carcinogenic, however, the amount is so small, and the concentrations of possible carcinogens so minute that carbon black simply cannot be said to cause cancer during manufacture or use. Industry investigations have shown cancer incidence to be no higher among former carbon black plant employees than in comparable other groups.

III. REINFORCEMENT

The dramatic increase in properties like modulus, hardness, tear strength, and abrasion resistance that appears when carbon black is added to rubber has caused many researchers to try and find the reason for this reinforcement. Practice seems to outpace theory in this area. It may be that there are both physical and chemical interactions involved. By heating black above 1300°C and compounding in rubber, both modulus and abrasion resistance are reduced. Such treatment serves to devolatilize the black.

Some progress has been made in understanding the viscoelastic properties using the Guth and Gold equation [8]:

$$\frac{M_f}{M_g} = 1 + 2.5\phi + 14.1\phi$$

Where M_f is the modulus or viscosity of the filled compound, M_g the modulus or viscosity of the gum compound, and ϕ the volume fraction of particles. This is an extension of the theory developed by Einstein [9] for the increase in viscosity of a liquid due to the addition of spherical particles, which takes into account particle interactions that occur at higher concentrations. In practice, reinforcing blacks give higher modulus than is predicted by the formula. The reason is that carbon blacks are aggregates, not spheres. When provision is made for the rubber that is within the interstices of the aggregate, better predictions can be made.

Hysteresis plays an important part in rubber compound behavior. If measured as the fraction of mechanical energy input that is converted to heat in a cyclic strain (tan δ), it relates to surface area and volume fraction. In general, blacks with better tread wear characteristics are higher in hysteresis. A review of the mechanism of reinforcement is given by Medalia et al. [10]. The literature of reinforcement includes the works of Payne [11], Lake and Thomas [12], Bueche [13], and Dannenberg [14].

The whole subject of carbon black reinforcement may enter a new era with the use of fractal mathematics. A seminal paper on this may be that of Gerspacher: "The Role of Carbon Black in Compound Viscoelastic Properties [15].

IV. MANUFACTURE

Five types of carbon black, based on manufacturing method, have been or are used in rubber compounds. Roughly in order of their development they are:

1. Lampblack
2. Channel black
3. Thermal black
4. Acetylene black
5. Oil furnace black

By far the most popular blacks are those made by the oil furnace process. A small amount of lampblack is used in Europe. Channel black, a very reinforcing black, was made and exported from Romania until the civil disturbances recently; details of present production and usage are not clear. Acetylene black is a unique black made by the continuous thermal pyrolysis of acetylene above 1500°C in the absence of air. One of the earliest producers, if not the original, was the Shawinigan Chemical Company of Canada. That black, Shawinigan acetylene black, is now made at the Cedar Bayou, Texas, plant of Chevron

Chemical. This black is used in rubber because its structure imparts electrical and thermal conductivity. (Shawinigan was the original "structure" black against which others were judged.) Besides affording high electrical and thermal conductivity, it is a relatively clean black with less than 0.05% ash due to its feedstock. It has been estimated that the total length of chain conductor in acetylene black is 3.9×10^5 km/g.

Thermal blacks, made from natural gas, were solely an American development, and commercial production started in about 1922 at the Thermatomic Carbon Company in Sterlington, Louisiana. As the price of natural gas increased, thermal carbon prices followed, and competitive materials took over much of the market. Thermal plants in the United States shut down, and production shifted to Alberta, Canada, where natural gas supplies are plentiful.

A. Thermal Blacks

Thermal blacks are made by the thermal decomposition of natural gas in a preheated vertical furnace with a refractory brick lining and filled with a checkerboard pattern of silica brick. Furnaces are perhaps 10–15 ft in diameter and 40 ft high. Two furnaces in tandem constitute a unit. One furnace burns natural gas and the stoichiometric amount of air for combustion for 5 min. This is the heating cycle. The air is then turned off and natural gas alone or with hydrogen is passed through for 5 min. The carbon black and hydrogen derived from the natural gas are swept along to spray towers, where water sprays cool the mixture and the black is collected in bag filters. If natural gas alone is used in the make cycle, the product is medium thermal (MT) black. After collection in the bag filters, the fluffy black is passed through a magnetic separator and a micropulverizer, which removes any tramp iron and breaks up lumps, respectively. The black can then be bagged as fluffy black or, more likely, be pelletized and shipped in bags or tank cars. Temperatures in the furnace reach about 1300°C. By operating one furnace on heating while the other is making black and then reversing, a continuous process is achieved. By using cycles shorter than 5 min, a semireinforcing furnace black can be made by this process. A drawback is the much more rapid disintegration of the interior brick with consequently higher grit levels in the black. Fine thermal (FT) black formerly was obtained when natural gas was used with hydrogen, the hydrogen being recycled. This product, marketed as P-33, was used in inner tubes. With the introduction of tubeless tires, and certain later developments, demand for FT black decreased, and it is no longer manufactured.

B. Channel Blacks

Channel blacks produced for rubber compounding in the United States were another black casualty due in part to rising natural gas prices but also to the introduction of oil furnace blacks n the middle 1940s. The latter were found to be better suited to synthetic rubber, which was then being used in increasing volume. Although use of rubber grade channel blacks in the United States is negligible, such grades have beer produced in Romania and Indonesia.

In the channel process thousands of small, natural gas flames impinge on iron channels, which move slowly back and forth as scrapers remove the accumulated carbon. The burners and channels are housed in low buildings called hothouses, so constructed that limited supplies of air reach the burners resulting in soot-laden flames. The yield—about 1.5 b/1000 ft³ of natural gas—was the lowest of any black process. This, coupled with the polluting atmosphere it produced, made its production unacceptable in many areas.

C. Oil Furnace Blacks

By far the most important carbon black process is the oil furnace process. First developed by the Phillips Company in the early part of World War II, it utilized petroleum feedstocks rather than natural gas, was a high-capacity process, and recovered much more carbon from the feed material. Other companies, such as Cabot, developed the method further by altering how the combustion air was fed, burner design, etc. Freed from dependence on natural gas, the industry could now be located wherever appropriate feedstocks could be shipped, and a world-wide industry was created.

Feedstocks for this process are highly aromatic, highly viscous materials (with high C/H ratio) at the lower end of refiners' distillation ranges. Strict feedstock control is vital; alkali metals, sulfur, and ash contents must be carefully limited, with only very low levels of asphaltenes and solid carbon particles permitted.

In a typical process the feedstock is preheated, perhaps in a liquid salt bath and then, atomized with air, is squirted down the center of the refractory-lined horizontal reactor. Air and fuel gas introduced tangentially into the reactor at the same end burn and form an enveloping ring around the oil-rich aerosol in the center. Under these circumstances the feedstock droplets decrease in size and increase in carbon content until they are almost all carbon. There are numerous collisions between these particles, and aggregates are formed. A quench water spray near the exit end of the reactor effectively stops the reaction. Alkali metal salts may also be injected into the reactor to modify the modulus and structure of the black if needed.

From the reactor, the carbon-laden gases are cooled in water spray towers. The black is separated from the gas stream by cyclone separators and bag filters. The bags are usually made of glass or Teflon. Hard, dense particles are screened out and sent through a micropulverizer before being returned to the product stream. When the black is sufficiently densified, it is pelletized in cylindrical horizontal tanks, agitated by a central shaft with a helix of metal pins attached to it. Water, possibly with a binder such as molasses, is added to the black as it enters the pelletizer.

The black emerges as moist pellets, which are dried in a rotating drum with a countercurrent of hot gas, passed over a magnetic separator to remove any tramp iron, screened, and bagged or held in vertical storage tanks to provide tank car shipments. Pellets are dried to less than 1% moisture. The gases exiting the process have a low calorific value but can be burned to heat utility water, etc.

Present-day carbon black production is tightly controlled with computer simulations of the process, and computer and microprocessors used for flow controls of inputs. These measures have sharply reduced if not eliminated problems the compounder had in the past with carbon black shipments. These included erratic or slow-curing lots due to a change in feedstock, shipments with impaired electric resistance in wire cable stocks due to high mineral salt concentration in the black from cooling and pelletizing water, and off-quality black due to loading while still hot into tank cars or into bags, with resultant seam opening.

V. GRADING CARBON BLACKS

As the range of carbon blacks grew, the nomenclature expanded, with acronyms indicating manufacturing process, fineness of the particle, processing quality in rubber, etc. Channel blacks were identified as EPC, MPC, HPC, the letters EP, MP, and HP standing for easy, medium, or hard processing, and C for the channel process. For the medium and fine thermal blacks (MT and FT), the first letter identified the size of the average particle, relatively speaking, with T indicating a thermal process. Gas furnace blacks had labels SRF (semireinforcing furnace) and HMF (high modulus furnace), where "semireinforcing referred to the inferior abrasion resistance compared to channel blacks and "high modulus" to the high stress needed to elongate the vulcanizate containing it. The introduction of oil furnace blacks broadened the nomenclature further. The first such black, Philblack A from Phillips, was identified as FEF (fast-extruding furnace). Later blacks usually were designated by their relative abrasion, and HAF, ISAF, and SAF meant, respectively, high, intermediate, and superabrasion furnace black.

This mushrooming of categories was channeled and controlled by the ASTM in the classification system they adopted for carbon blacks used in rubber products, D1765 [16]. This scheme uses an initial letter S or N followed by three digits. N indicates that the black in rubber has the normal curing rate of furnace blacks; S means that curing is slow, typical of channel blacks or furnace blacks that have been modified. The next digit indicates the typical average particle size from various suppliers covering the 1- to 500-nm range in 10 grades (not in equal ranges). The two digits that follow are arbitrarily chosen. The list changes, of course, as new blacks are added and old ones deleted. The grades are well bracketed by the number of typical properties given. As measures of surface area they give iodine adsorption numbers, CTAB, and nitrogen adsorption values. Tinting strength values are listed. As mentioned earlier, this test measures the light absorption coefficient and is related primarily to particle size but also depends on structure and aggregate size distribution. As indicators of structure, dibutyl phthalate absorption numbers are given for both uncompressed and compressed samples. Pour density is included, as is the difference in 300% modulus at two cure times from a reference black when specimens from both times are tested and compared in a natural rubber formula. In addition, all blacks must have a screen residue less than 0.0010% when tested on a 500-μm (no. 35) screen, less than 0.10% on a 45-μm (no. 325) screen, ASTM method D1514. All blacks are required to have a maximum of 7% fines on screen size 125 μm (no. 120) except thermal blacks when tested by ASTM method D1508.

VI. COMPOUNDING

A. Introduction

Black is received at the plant in buk (tank cars or tote bags), in paper bags, or in masterbatch form. Bagged black should be stored under cover in a dry place to minimize moisture pickup. Bulk back is stored in large cylindrical vertical steel tanks. Such tanks and auxiliary conveying equipment should be routinely examined to make sure that no rust and/or scale is getting into the black.

As noted previously, carbon black producers have instituted so many controls that extensive incoming sampling and testing is unnecessary. Routine checks for moisture, pellet size, grit, and pour density are quick and give some insurance against difficulty in processing. Less frequent spot checks that are useful include iodine surface area, dibutyl phthalate absorption, and the vulcanizate properties in the NR test formula ASTM D3192. Comparison of interlaboratory results has been made more accurate by determining the differences in rubber properties of the black in question from those obtained with a standard industry reference black (IRB).

B. Mixing

Mixing of blacks into rubber still smacks more of an art than a science. Significantly different procedures are used by established companies to achieve acceptable results. This means that the pellets are crushed and brought to a satisfactory dispersion in the mix with a minimum of mixing time, energy, and heat buildup. Obviously some compromises have to be made. Usually half if not all of the black is added early in the mixing cycle, with the result that the high shearing forces during polymer breakdown crush the pellets and scatter the agglomerates thoroughly. It is rare too that a softener or plasticizer is not added to facilitate black dispersion.

It is hard to set limits on the capacity of a rubber mix to accept black before mixing becomes so difficult that the compound is unworkable. Governing factors include polymer used, type of black, softener content, mixer employed, and whether the black is in masterbatch form. Trade literature [1] indicates, for example, that SBR 1500 with 70 parts of ISAF black and 10 parts of oil (softener) has a Mooney viscosity of 84 (4 ML at 212°F). A 70-part loading of ISAF black in NR with 3 parts of oil has a Mooney value so high it cannot be measured, but if HAF black is used the Mooney is 83. For much rubber equipment such high viscosities would cause trouble, perhaps from premature cure or scorch, as the condition is called. As a first approximation, loadings that give a Mooney viscosity exceeding 80 might be considered to be unacceptable. Frequently masterbatches of commonly used blacks are made with natural rubber or SBR to facilitate later mixing.

C. Black Selection

Two properties of black relate strongly to its reinforcing ability: surface area and structure. As surface area increases, the viscosity of the uncured stock increases: mixing takes longer, and the smoothness of extrusions increases. In the vulcanizate, abrasion resistance, tear strength, cut growth, and flex resistance all increase, along with tensile strength and modulus. Increasing structure decreases tensile strength but increases modulus, hardness, and abrasion resistance.

In the 1990 ASTM classification for carbon blacks, no less than 42 commercial blacks were recognized, as well as the current reference black, IRB 6. As one lot of industry reference black becomes exhausted, another lot is selected, tested, and given the next number. There is much overlap among the 42 blacks with respect to the characteristics they impart to rubber. For most rubber goods companies, six blacks will suffice, along with a supply of the latest IRB. The larger the number of blacks carried, the higher the inventory charges and the greater the chances of mixups. The six blacks deemed sufficient might well vary depending on what products are made. For a mechanical rubber goods producer, the choice, identified by ASTM number and common industry reference term, might be N293 conductive furnace, N220 intermediate superabrasion furnace, N330 high abrasion furnace, N550 fast-extruding furnace, N660 general-purpose furnace, and N990 medium thermal. It is understood of course that blends of two blacks are often used. In choosing these workhorse blacks, the grades should be popular enough and useful enough that they will not be withdrawn from the market completely.

Table 10.1 gives typical analytical values for iodine absorption number, dibutyl phthalate number, and pour density for the foregoing blacks, as abstracted from ASTM D1765, along with the maximum heat loss from a survey of the literature. The iodine absorption numbers, which approximate the surface area in square meters per gram, cover the range given in ASTM D1765. The

Table 10.1 Analytical Properties of Selected Blacks

ASTM no.	Iodine adsorption (g/kg)	DBP no. (cm³/100 g)	Pour density (lb/ft³)	Heat loss (% max)
N220	121	114	21.5	3.0
N293	145	100	23.5	2.5
N330	82	102	23.5	2.5
N550	43	12	22.5	1.5
N660	36	90	26.5	1.0
N990	6–12	43	40.0	0.10
IBR 6	80	100	24.2	2.5

DBP number range is not so extensive, covering only about 62% of the range. DBP no. is the *n*-dibutyl phthalate absorption number of the compressed sample. This should not cause problems. Some of the blacks outside the selected range are definitely specialty blacks such as N472. The heat loss values are given as a reminder that carbon black is hygroscopic, and careful attention should be given to ensure that these values are not exceeded.

From the foregoing information, the amount of black that can be used is estimated roughly. Recourse to suppliers' literature helps the compounder to zero in on the type of black most suitable and the amount for the desired compound. Most carbon black producers supply data on their blacks at a 50-part loading in various elastomers and often data at other loadings as well. Several properties for carbon blacks of a major producer (Cabot) in six of the elastomers reviewed in this book are abstracted from the company's published figures [17] and given as Tables 10.2–10.7. Since, as noted earlier, polybutadiene is rarely compounded alone, data on this black are omitted. Because of the specialized use of poly-suflide rubber, comparative data on those materials would be of little use. Although Cabot has ceased production of MT black, the values given should be representative of continuing producers' quality. From such data a preliminary choice of the type of black can often be made to match the requirements of a given compound. High tensile strength requirements will be met by blacks like N220 or N293; if top resilience is wanted, an N990 black would be preferred.

The amount of the chosen black can often be estimated from producers' tables as well. Table 10.8 illustrates how changing the loading varies the rubber properties. These values, from a much larger extensive study published by Phillips [1], show the change in properties when an HAF black is used in an SBR 1500 formula at 30-, 50-, and 70-part loadings. Certain characteristics of increasing black content become apparent even in this abbreviated table. Note how rebound resilience and compression set decline as loading is increased.

Table 10.2 Properties of Selected Blacks in a Natural Rubber[a]

ASTM no.	Industry type	300% Modulus (psi)	Tensile strength (psi)	Elongation at break (%)	Hardness, Shore A	Rebound resilience (%)
N220	ISAF	1770	3720	550	61	60.7
N293	CF	1235	3621	610	60	59.0
N330	HAF	1700	3660	545	60	65.1
N550	FEF	1950	3256	485	58	71.6
N660	GPF	1666	3310	530	57	72.9
N990	MT[b]	625	3025	620	49	84.0

[a]Formula (cured at 320°F, 160°C)

SMR 5CV	100.0
Carbon black	50.0
Zinc oxide	5.0
Stearic acid	3.0
Altax[c]	0.6
Sulfur	2.5
	161.1

[b]No longer in production at Cabot Corporation.
[c]Benzothiazyldisulfide. (Vanderbilt).
Source: Technical Bulletin 3G(1991), Cabot Corporation, Boston, Massachusetts.

Table 10.3 Properties of Selected Blacks in an SBR 1500 Rubber[a]

ASTM no.	Industry type	300% Modulus (psi)	Tensile strength (psi)	Elongation at break (%)	Hardness, Shore A	Rebound resilience (%)
N220	ISAF	2540	4332	445	70	44.2
N293	CF	2157	4241	500	68	54.7
N330	HAF	2433	4230	470	68	59.0
N550	FEF	2306	3472	435	66	65.3
N660	GPF	1965	3420	504	63	66.4
N990	MT[b]	600	1650	580	53	70.0

[a]Formula (cured at 293°F, 145°C)

SBR 1500	100.0
Carbon black	50.0
Zinc oxide	2.0
Stearic acid	1.0
TBBS[c]	1.0
Sulfur	1.8
	155.8

[b]No longer in production at Cabot Corporation.
[c]N-tert-Butyl-2-benzothiazolesulfenamide.
Source: Technical Bulletin 3G(1991), Cabot Corporation, Boston, Massachusetts.

Table 10.4 Properties of Selected Blacks in a Neoprene W[a]

ASTM no.	Industry type	300% Modulus (psi)	Tensile strength (psi)	Elongation at break (%)	Hardness, Shore A	Rebound resilience (%)
N220	ISAF	2204	3240	284	76	49.3
N293	CF	1734	3124	330	73	48.4
N330	HAF	2063	2940	271	72	55.9
N550	FEF	1847	2862	309	70	61.4
N660	GPF	1484	2834	348	65	64.5
N990	MT[b]	380	2125	720	53	71.2

[a]Formula (cured at 320°F, 160°C)

Neoprene W[c]	100.0
Carbon black	50.0
Sundex 790[d]	12.0
Stearic acid	0.5
Magnesium oxide	8.0
Antioxidant	2.0
Antiozonant	1.0
TMTD[e]	0.5

for every rubber except EPDM, where the relationship is about 2.3.

[f]Ethylene thiourea.

Source: Technical Bulletin 3G(1991), Cabot Corporation, Boston, Massachusetts

f the black to
mated rather
s of the other
for about 2.1
cidentally, it
by one point
.3.
compound or
ss is usually
tock contain-
s: to natural
.8 (10/2.1) to
e read closer
ations are for

Some compound specifications require certain hardness range
be used is an N330 (HAF) black, the amount required can be
accurately. It is known that with the exception of neoprene, hard
polymers studied here will increase by one Shore A hardness u
parts of black added. For neoprene W it is about 1.7 parts. C
takes about 2.1 parts of common process oils to reduce the hardn
for every rubber except EPDM, where the relationship is abo

Also required is the hardness of the rubber as a pure gu
hardness as black loading is projected to zero. Pure gum har
around 41. We can then estimate the hardness of a natural rubbe
ing 60 parts of HAF black and 10 parts of process oil as fol
rubber's pure gum hardness of 41 add 28.6 (60/2.1), then subtrac
get 64.8. Since it is doubtful that Shore A durometer gages ca
than one unit, the hardness would be estimated as 65. These cal

Table 10.5 Properties of Selected Blacks in a Butyl[a]

ASTM no.	Industry type	300% Modulus (psi)	Tensile strength (psi)	Elongation at break (%)	Hardness, Shore A	Rebound resilience (%)
N220	ISAF	982	2570	605	63	29.6
N293	CF	748	2595	650	60	29.1
N330	HAF	917	2426	612	62	30.8
N550	FEF	912	1823	578	58	34.1
N660	GPF	775	1813	597	56	33.9
N990	MT[b]	300	2200	700	52	40.0

[a]Formula (cured at 320°F, 160°C)

Butyl 268[c]	100.0
Carbon black	50.0
Sunpar 115[d]	5.0
Zinc oxide	5.0
Stearic acid	1.0
Ethyl cadmate[e]	2.0
MBTS[f]	0.5
Sulfur	1.0
	164.5

[b]No longer in production at Cabot Corporation.
[c]From Exxon Chemical Americas, Houston.
[d]Paraffinic process oil (Sun Refining and Marketing Co.).
[e]Cadium diethyldithiocarbamate.
[f]Benzothiazyldisulfide.
Source: Technical Bulletin 3G(1991), Cabot Corporation, Boston, Massachusetts.

an HAF black; large particle blacks increase hardness much less for the same loading. It takes 5.2 parts of a medium thermal black to raise hardness one unit. This technique also can be used with other fillers such as clay or whiting.

Snow and Barlow [18] studied some aspects of the curing rate of 21 carbon blacks in SBR, using five methods of determination. By one commonly adopted method (time to 90% cure as measured with the Monsanto rheometer), the spread was found to be from 34.0 min for superior processing furnace (SPF) black to 46.5 min for easy processing channel (EPC) black. Since this difference is significant, compounders should confirm that the black they have chosen will match the curing schedules needed. Finally, one of the more comprehensive studies on the mixing of carbon black in rubber was given by Hess et al. [19]. We now turn to the blacks developed roughly in the period 1972–1977 whose physical properties were referred to earlier. Typical of these blacks are those bearing ASTM numbers N234, N339, N351, and N375. By far the greatest volume of these blacks is used in tire treads, especially passenger tire treads, where only a single black type is used. Attractive to tire makers is that increased

Table 10.6 Properties of Selected Blacks in an EPDM[a] at 100-part Black Loading

ASTM no.	Industry type	300% Modulus (psi)	Tensile strength (psi)	Elongation at break (%)	Hardness, Shore A	Rebound resilience (%)
N220	ISAF	1126	2612	420	70	48.6
N293	CF	610	2297	552	67	49.0
N330	HAF	1290	2370	355	68	51.1
N550	FEF	1291	2280	415	66	57.3
N660	GPF	1022	2130	462	62	61.6
N990	MT[b]	250	830	660	48	66.0

[a]Formula (cured at 310°F, 154°C)

Royalene 502[c]	100.0
Carbon black	100.0
Sunpar 2280[d]	50.0
Zinc oxide	5.0
Stearic acid	1.0
MBT[e]	0.3
TMTM[f]	1.0
Sulfur	1.0
	258.3

[b]No longer in production at Cabot Corporation.
[c]From Uniroyal Chemical, Middlebury, CT.
[d]Paraffinic process oil (Sun Refining and Marketing Co.).
[e]Mercaptobenzothiazole.
[f]Tetramethylthiurammonosulfide.
Source: Technical Bulletin 3G(1991), Cabot Corporation, Boston, Massachusetts.

mileage is obtained with little or no increase in black cost. In mechanical goods these blacks may be used where the ultimate in abrasion resistance is wanted—in heavy-duty conveyor belts, for instance. Compounding with these blacks is similar to compounding with conventional blacks of the same iodine surface area. There is a tendency to faster curing, so scorching tendencies should be carefully noted.

Medalia et al. [10] suggest that the improved wear is due not only to higher surface area (N_2 absorption) but to the following considerations:

1. Higher surface activity.
2. Narrower distribution of particle size and visual absence of very large particles, which are believed to act as weak spots in abrasion or tensile failure.
3. More open aggregates.

Few firm rules can be given for the selection and dosage of carbon blacks for tire treads, especially passenger treads. At present compounders appear to be

Table 10.7 Properties of Selected Blacks in a Medium Acrylonitrile[a]

ASTM no.	Industry type	300% Modulus (psi)	Tensile strength (psi)	Elongation at break (%)	Hardness, Shore A	Rebound resilience (%)
N220	ISAF	1580	3180	490	55	53.8
N293	CF	1235	3227	571	54	52.6
N330	HAF	1640	2820	450	54	57.0
N550	FEF	1432	2540	490	51	61.8
N660	GPF	1140	2350	550	48	63.1
N990	MT[b]	420	2100	900	45	65.0

[a]Formula (cured at 320°F, 160°C)

Hycar VT 380[c]	100.0
Carbon black	50.0
Dioctyl phthalate	25.0
Zinc oxide	3.0
Stearic acid	1.0
Antioxidant	1.0
MBTS[d]	1.5
TMTM[e]	0.2
Sulfur	1.5
	183.2

[b]No longer in production at Cabot Corporation.
[c]From Nippon Zeon, Cleveland, OH.
[d]Benzothiazoledisulfide.
[e]Tetrathiurammonosulfide.
Source: Technical Bulletin 3G(1991), Cabot Corporation, Boston, Massachusetts.

under the gun to provide compounds meeting the automotive industry's demands for high wet skid resistance, low rolling resistance (to save energy), high mileage, and other qualities, some of which are incompatible. Predicting when this situation will stabilize is very speculative.

Before leaving this compounding section, some mention should be made of formulating electrically conductive rubber stocks. Invariably these compounds achieve their conductivity through the use of carbon blacks, either acetylene black or conductive oil furnace black. Electrical resistivity—that is, volume resistivity—is measured by such methods as ASTM D991-89 and may be expressed in ohm-centimeters ($\Omega \cdot cm$). The log of resistivity of rubber compounds might be 8 to 10 for general-purpose compounds, 4 to 8 for antistatic compounds, and under 4 for conductive compounds. Antistatic compounds are used in such application as flooring or sheeting in hospitals, to prevent static electricity discharges from igniting ether–air mixtures in operating rooms. Conductive rubber compounds have been used in underground mining cable covers and/or interior layers to protect miners whose metal tools might cut through to

Table 10.8 Loading Study of an HAF Black in SBR 1500

Property			Compound no.		
			2	6	10
4 ML at 212°F			35	50	74
Scorch 280°F, min			27	21	16
	Stress–Strain (at room temperature)				
	Cure at 307°F (min.)				
300% modulus,	30		750	1580	2550
psi	45		740	1650	2750
Ultimate tensile,	30		2700	2960	2960
psi	45		2625	3010	3080
Elongation at	30		590	505	350
break, %	45		615	505	340
Hardness, Shore A	45		53	66	77
	Stress–Strain (at 212°F)				
300% modulus, psi	30		540	1340	1450
Ultimate tensile, psi	30		675	1430	1450
Elongation at break, %	30		345	380	

Mooney Data

140

	Stress–Strain (aged 70 hr at 212°F in air oven)			
300% modulus, psi	30	1480		
Ultimate tensile, psi	30	2090	2610	2760
Elongation at break, %	30	340	285	180
Shore A hardness	30	60	71	82
Special Data				
Heat buildup, °F	45	47	58	76
Rebound resilience, %	45	67.1	58.7	50.9
Compression set, %	30	28.5	27.8	20.4
Component				
SBR 1500		100	100	100
HAF black (Philblack 0)		30	50	70
Zinc oxide		5	5	5
Thermoflex A[a]		1	1	1
Stearic acid		2	2	2
Circo Paraflux[b]		10	10	10
Santocure[c]		1.10	1.05	1.05
Sulfur		1.75	1.75	1.75
		150.85	170.80	190.80

[a]Antioxidant.
[b]Process oil (Sun Oil Co.).
[c]N-Cyclohexyl-2-benzothiazolesulfenamide (Monsanto).
Source: Technical Bulletin 3G(1991), Cabot Corporation, Boston, Massachusetts.

the conducting wire; and such compounds also have been used in flexible heating pads.

The amount of carbon black used in these applications is small. For example, worldwide acetylene black product capacity in 1983 was estimated at less than 0.5% of estimated total capacity, and probably more than half of this went into dry cells. Production of conductive oil furnace black is difficult to estimate. At one time, one carbon black company simply bagged one of the plant's production as ISAF or CF (for conductive furnace black), depending on what electrical conductivity the product seemed to be giving at the time.

What makes conductive rubber compounding worthy of mention is because of some basic differences from conventional compounding. Blacks used to impart conductivity are few. Shawinigan acetylene black is a common brand in the United States. Among the oil furnace blacks, ASTM N293 is frequently used. Low loadings of black (say 20–30 parts) impart little conductivity; and loadings as high as 50–80 parts are usually required to get stocks with conductivities of 30 or less (log of the resistivity). Resistivity is a function of surface area (the lower the surface area, the greater the resistance). Thus high surface area blacks (> 100 m^2/g: N_2 adsorption), for instance, are preferred.

Again higher structure leads to greater conductivity. The accepted theory is that with high surface area, high structure blacks there are strings of carbon particles, largely in contact, through the rubber matrix to provide conductive paths Carbon blacks should be chosen that have little extractable tarry material on them that would insulate the particles from each other in the chain.

Polymers differ in conductivity. EPDM, for example, is relatively conductive, natural rubber moderately so, and SBR more insulating. Usually other factors than electrical conductivity determine the choice of polymer. One most important consideration in conductive stock production is mixing. Unlike other compounds, excellent dispersion of the black is not a goal. The objective is to disperse the black sufficiently to permit the adequate development of such properties as modulus and hardness, yet there is enough undispersed black left in the threads or chains throughout the system to have the conductivity desired.

Two further difficulties arise in obtaining satisfactory conductive stocks. Flexing the cured rubber piece decreases the conductivity significantly, and not all of this property is recovered after the distortion has ceased. The increase in resistance is affected by the type of black used, the polymer chosen, and the amount of flexing that has occurred, among other factors. Second, electrical conductivity measurements of rubber pieces vary widely, and close adherence to a detailed procedure is essential for reproducible results.

D. Ground Coal

A relatively new compounding ingredient, ground bituminous coal, is not a reinforcing pigment but is included here because it is essentially carbon. Ground coal is used primarily as a low cost extending filler. At the time of writing the cost was approximately one-third that of carbon black, although several times that of the clays and whitings described later. Ground coals may have 75% fixed carbon, 18% volatiles, and 6–7% ash. The volatile content does not evolve at mixing or processing temperatures. A proposed ASTM specification lists three grades of this material with specific gravities ranging from 1.35 to 1.55. Irrespective of this difference, the specific gravity is considerably lower than that of carbon black (1.80) and clays and whitings (2.60–2.70). A descriptive leaflet by one supplier states that 99.5% passes through a 500 mesh screen. This is a very fine material, but when tested in an explosion chamber, no explosion could be created. Basic information is given in Trojan and Klingensmith [20].

REFERENCES

1. *Philblack in Natural and Synthetic Rubbers*, Bulletin P-21, Phillips Chemical Co., Bartlesville, OK, 1960.
2. Lyon, F., and Burgess, K. A., *Encyclopedia of Polymer Science and Engineering*, Vol. 2, 2nd ed., p. 627, (1985).
3. *1991 Annual Book of ASTM Standards*, Vol. 09.01, p. 270.
4. *1991 Annual Book of ASTM Standards*, Vol. 09.01, p. 620.
5. *1991 Annual Book of ASTM Standards*, Vol. 09.01, p. 438.
6. Lyon, F., and Burgess, K.A., *Encyclopedia of Polymer Science and Engineering*, Vol. 2, 2nd ed., p. 633.
7. Swor, R. A., Hess, W. M., and Micek, E. J., *New High Structure Tread Black for Compounding Flexibility*, presented at the 138th Meeting, of the American Chemical Society, Rubber Division, Washington, DC, October 1990.
8. (a) Guth, E., and Gold, O., *Phys. Rev.*, *53*, 322 (1938). (b) Guth, E., *Proc. 5th Int. Cong. Appl. Mechanics*, Cambridge, 1938, p. 448.
9. Einstein, A., *Ann. Phys.*, *19*, 289 (1906); *34*, 591 (1911).
10. Medalia, A. I., Juengel, R. R., and Collins, J. M., in *Developments in Rubber Technology*, A. Whelan and K. S. Lee, eds., Eng. Applied Science Pub. Ltd., Essex, 1979, pp. 151–181.
11. Payne, A. R., in *Reinforcement of Elastomers*, G. Kraus, ed., Wiley. Interscience, New York, 1965, Ch. 3.
12. Lake, G. J., and Thomas, A. G., *Proc. R. Soc.*, *A300*, 108 (1967).
13. Bueche, F., *Physical Properties of Polymers*, Wiley, New York, 1982.
14. Dannenberg, E. M., *Trans. Inst. Rubber Ind.*, *42*, T26 (1966).
15. Gerspacher, M., The role of carbon black in Compound Viscoelastic Properties, presented at the 140th Meeting, of the American Chemical Society, Rubber Division, Detroit, 1991.
16. *1991 Annual Book of ASTM Standards*, Vol. 09.01, p. 438.
17. Philblack on Natural and Synthetic Rubbers, Report RG-136, Cabot Corp., Boston, MA.
18. Snow, C. W. and Barlow, F. W., *Curing Rates of Carbon Blacks in SBR by Classical and Modern Methods*. Presented at the meeting of the Rubber Division of the American Chemical Society, New York, September, 1966.

19. Hess, W. M., Swor, R. A., and Micek, E. J., *Influence of Carbon Black Properties on Dispersion*, presented at the Fall meeting The Rubber Division of the American Chemical Society Houston, October, 1983.
20. Trojan, M. J., and Klingensmith, W., New Views on Using Ground Coal Fillers, Rubber Division Meeting, ACS, Mexico City Meeting, May 1989.

11

Nonblack Fillers

I. INTRODUCTION

In classifying the dry fillers used in rubber to increase its usefulness or to make a cost-competitive product, it is convenient to speak of black and nonblack fillers. With few exceptions (such as color pigments) nonblack fillers are chosen over carbon blacks or ground coal for one or more of three reasons:

1. The product has to be white or light-colored.
2. Certain unique properties are required like thermal conductivity by the use of zinc oxide.
3. Cost—it is hard to get less costly materials than natural products in good supply like clay or ground limestone.

On a weight basis, about 30% of the rubber fillers used in the United States are inorganic nonblack pigments. From an energy conservation standpoint, it would be helpful if a greater proportion could be used. It has been estimated that producing an ISAF black requires 90–110 MJ/kg compared to 70 for precipitated silica and less than 0.5 for air-floated clays [1]. The energy consumed in making carbon blacks comes almost entirely from petroleum sources. Such dependence comes against a backdrop of new oil fields that seem to be depleting more quickly than expected. Some carbon blacks of good quality are made from coal tar products (such as anthracene residues), but the supply of such feedstocks is limited. Some progress in heat conservation is being made in carbon black plants.

Nonblack fillers have certain features in common compared to carbon blacks. They have higher specific gravities—from 1.95 for a precipitated hydrated silica to 5.6 for zinc oxide. At the same oading by weight, they ordinarily have lower tensile values than blacks. For example, in a 50-part loaded nitrile stock with 25 parts of plasticizer, the highest tensile strength for the most reinforcing nonblack filler, silica, would probably be 1800–2000 psi; carbon blacks would range from 1900 to 3000. There is simply not the chemical and physical interaction between filler and polymer that exists with carbon black. For the same reason, modulus is lower at the same hardness. Inasmuch as high tensile strength and modulus are considered to be vital components of abrasion resistance, nonblack fillers are less abrasion-resistant.

Certain differences show up in process as well. There is more chance that natural nonblack fillers, the most popular of the nonblack group, will have more oversize particles (i.e., material not passing a 325-mesh screen) than blacks. Oversize particles, which may be points of initial rupture in a given stock, can easily lower tensile or tear strength. Accompanying such grittiness is the scoring and more rapid wear of parts like ruber dies. Although carbon blacks (except thermal) are aggregates of particles, most nonblack fillers are not; they consist of acicular, platey, or blocky particles. This difference can show up in shrinkage with or across the grain in calendered or extruded stocks, and such shrinkage cannot be predicted from carbon black stock characteristics.

Because of the inferior reinforcing power of nonblack pigments as produced, much recent work has been aimed at treating them to improve their properties. Some successes have been obtained and are mentioned later.

There are many ways to classify nonblack fillers considering the variety of natural sources and various treatments now being used. We shall discuss three, identified by their basic compositions: clay, silica and silicates, and calcium carbonates. These three sources, however, constitute in their as-produced and treated forms, over 90% of the nonblack fillers used in the United States.

It has been estimated that 54% of the U.S. rubber industry's consumption of nonblack fillers is clay, 27% calcium carbonate, 15% silicas and silicates, and 4% other materials [2]. There is an economic reason for these differences [2]. Clays vary in price from 0.03 to 0.14 $/lb, depending on treatment; calcium carbonates cost 0.02–0.20 $/lb, depending on quality; and the cost of precipitated and fumed (anhydrous) silica runs between $0.45 and $2.10/lb, depending largely on method of manufacturing. Although reference has been made to zinc oxide as a nonblack filler, its cost limits its use mainly to such applications as activator for curing systems and covulcanizing agent with magnesium oxide in neoprene compounds. Its properties were reviewed in Chapter 7.

II. CLAY

One of the most widely used nonblack fillers for rubbers is clay. Its use is based on its low comparative cost, versatility, and stiffening properties. Large deposits, suitable for rubber, were found and developed in England and the United States. Strictly speaking, clay refers to a physical condition, not a chemical composition, but old calling habits die hard. Kaolin clay, the type of clay used in rubber, has been derived from the weathering of aluminous minerals such as mica and feldspar. The closest approach to its chemical composition would probably be $Al_2O_3 \cdot 2SiO_2 \cdot 2H_2O$.

Rubber grade clays are white to cream powders with a density of 2.60. Clay is not hygroscopic; with a 1% maximum moisture content as packaged, there is little pickup of moisture even at high relative humidities. Significant differences in screen residue on a 325 mesh screen have been reported from clays mined in the same general area in Georgia. One supplier has a 0.8% maximum specification, another a 0.75 % maximum for one grade, but all other grades have less than 0.40%. Although both materials are kaolin clays, differences in silica and alumina content suggest that clays from Georgia and from Devon (England) might impart significantly different properties to rubber.

Clay is prepared in four different ways; it is air-floated, water-washed, calcined, and chemically modified. The method of manufacture changes material characteristics, including some of the properties in rubber.

The majority of the clay used by the rubber industry is air-floated. The crude clay, with perhaps 25% moisture content, is dried to less than 2% moisture in a rotary dryer. On discharge, the clay is pulverized. Air currents remove the smaller particles from the clay, and the coarser particles are rejected.

The water-washed process is more complex. The clay as mined is suspended in water by means of a process called blunging, which entails turbulent action and a dispersing agent. After partial settling has occurred, the top suspension is drawn off and fractionated (by hydrosettling, settling tanks, etc.). The fractions usually are diluted and reblunged, and the clays bleached by means of an agent that converts the insoluble ferric iron oxide to soluble ferrous compounds. The suspension is then filtered by plate filter presses or drum-type filters, and the filter cake is dried by rotary or tunnel dryers.

Calcined clay is more a specialty clay, used mostly in the wire and cable industries for the good wet and dry electrical properties it gives to insulation compounds.

An important recent development in rubber clay technology was the introduction by the J. M. Huber Company of clays chemically modified by the use of silanes with amino or mercapto pendant groups. Essentially these are water-washed hard (described later) clays, where the silane is chemically bonded to the kaolin silicate sheet by the hydrolyzed silane group and the pendant amino or mercapto group can then crosslink with the elastomer. The result is a clay with improved reinforcement, as indicated by higher modulus, tensile strength, and tear strength. Clays contain a small amount of crystalline quartz, and the alpha form is a carcinogen. Thus for health reasons, air-floated clays probably will be phased out in favor of wet-ground materials, which release less quartz into the air.

Air-floated and water-fractionated clays are further divided into hard and soft clays, where "hard" and "soft" refer to the behavior of the clays in rubber. Hard clays give more reinforced stocks, having higher stiffness in extrusion, coupled with higher modulus and tensile strength, and higher resistance to tears and abrasion. Soft clays extrude faster and cure more quickly. The differences, although significant, do not approach those displayed over the rubber grade carbon black spectrum. At 100 parts loading—on the low side for most clay-loaded compounds—hard clay in a natural rubber stock would probably have 150–250 psi higher 300% modulus than the soft clay, 300–400 psi higher tensile strength, and up to 50 lb/in. higher crescent tear values. Comparisons are made more difficult because of the difference in cure rates and the resultant uncertainty as to whether there is an equivalent state of cure.

The difference in reinforcement between hard and soft clays is linked to their average particle size. The hard clays have a median particle diameter of 0.3 μm compared to 1.1 μm for the soft varieties. Surface area, by the Brunauer–Emmett–Teller (BET) nitrogen adsorption method, is 22–26 m^2/g for many hard clays, 12–15 m^2/g for soft clays. Unlike carbon blacks, which are composed essentially of spherical particles clustered in aggregates, clays are hexagonal platelets which are very thin compared to their length and breadth. Since the determination of the thickness of the particle is difficult, the mean particle diameter is really the estimated spherical diameter of the particle derived from sedimentation methods of analysis. Air-floated clays are usually acidic with a pH of about 5, which can go up to neutral depending on the clay source.

Water-washed clays have advantages over the air-floated ones, and this is reflected in the price. By water fractionation the grit retained on a 325 mesh screen is greatly reduced, down to about one-third of its original value. This can mean that the wear on extrusion dies is reduced and they will last longer. In water washing the pH approaches neutrality and a much brighter pigment is obtained. This obviously is an advantage in light-colored stocks.

Calcined clays are rarely used in industrial rubber goods except in the wire and cable industry. Calcining removes most of the bound water, thus increasing the resistance of insulation compounds. As a result of their treatment, calcined clays have median particle diameters larger and surface areas appreciably smaller than either air-floated or water-washed clays. They also have a slightly higher density, 2.63, because of the loss of the water of hydration. Huber markets its chemically modified kaolin clays using NuCap, NuLok, and Polyfil WC as trademarks. The latter is also a calcined clay and is especially useful in wire and cable compounding.

As a filler, clay has several attractions for compounders. It is cheap, uniform, and always in good supply. It is versatile, it can be used with all the rubbers discussed here, and it blends well with other pigments. Whereas not much more than 60 parts of a reinforcing black can be mixed into rubbers without incurring difficulties in processing (unless large volumes of softener are used), clay loadings will range from 25 to over 300 parts, with significantly less softener being required than with blacks. Clay-loaded stocks have low shrinkage—an advantage in manufacturing precision parts. In neoprene stocks clay gives excellent tear resistance up to 80 volumes. In butyl, clays give tensile strengths equal to that of blacks with the exception of the more reinforcing blacks, and small amounts in inner tube stocks give smoother tubing compounds.

Certain characteristics of clay demand consideration in compounding if inferior vulcanizates are to be avoided. First, clay is an adsorptive material and can adsorb organic accelerators to such an extent that they do not function at high loadings of clay. In fact one of the routine quality control tests for clay has been the adsorption of DPG accelerator. There is also a wide variation in the adsorptive strength, hard clay having considerably higher adsorptive propensity than soft clays. Usually benzothiazolesulfide and sulfenamide accelerators work well, and the amount should be increased perhaps 20% over what would be used in a black stock. The adsorptive effects can be mitigated by using small amounts of triethanolamine or polyethylene glycols. These function by being preferentially adsorbed at the active sites on the silica layer of the clay. Not only must the slower curing rate of clays in general be considered, it also must be realized that hard clays are significantly slower curing than the soft clays.

Clays have little covering power in rubber: their index of refraction at 1.56 is similar to that of rubbers, which may vary from 1.51 to 1.56. This compares with high covering power titanium dioxide at 2.76. Hard clays are generally darker than soft clays and have a creamy undertone. If higher brightness is wanted, water-washed clays can be used. Among the nonblack pigments, clay stocks show somewhat less water absorption than compounds with other white pigments. Water absorption after immersion at 95.5°C was observed in the Huber laboratories, and results at two immersion times given in Table 11.1 [3]. Water absorption is a matter of concern in such items as dam seals, water main joint rings, and hot water hose.

An odd phenomenon involving the use of clays has been the color changes resulting from the reaction between certain antioxidants and clays. The theory is that organic diamines, which can be oxidized to semiquinone forms, undergo catalytic oxidation in the presence of certain types of clay and the oxidized groups are highly colored [4]. This reaction occurs not only in the rubber

Table 11.1 Wate Absorption (% volume increase)

	Time in water at 95.5°C	
Pigment	1 week	4 weeks
Suprex clay[a]	7.52	25.73
Paragon clay[b]	9.44	39.30
Hydrated aluminum silicate	9.92	38.55
Treated precipitated CaCO$_3$	13.24	43.89
Gilder's whiting	11.97	53.85
Ground limestone	9.91	37.59

[a]Hard clay.
[b]Soft clay.

compound; ''colored clay'' can develop by storing the clay and antioxidant in adjacent open bins. When a cornmon antioxidant—a reaction product of diphenylamine and acetone—is used, an uncured stock (clay-loaded) turns from its normal light tan color to green and produces a vulcanizate with tan-green undertone.

Since there appear to be no specific relationships that cover all antioxidants, the compounder will either check his mixes for this tendency to change color or get the information from his supplier. There can be wide spreads in tensile strength and cure rate of clay stocks depending on the accelerator, with CBS probably giving the highest tensile. Extrusion rate is also influenced by accelerator choice. DPG gives the highest extrusion rate, but CBS is also efficient here.

One of the reasons clays, silicas, and calcium carbonates give low stress–strain properties compared to carbon black is that a strengthening effect of the reaction between filler particles and polymer hardly exists. The nonblack fillers are hydrophilic, and carbon black, with its hydrophobic surface, is wetted more easily by elastomers. Another disadvantage has been the slow curing of clays. Both disadvantages have been attacked and with some success.

By treatment with bases and organic amines, clays that are faster curing than standard types have been prepared and marketed. But more important has been the development of clays using so-called coupling agents. These are difunctional organic compounds able to react first with the kaolin surface and then with the elastomer. Small amounts of aminosilanes and mercaptosilanes can be reacted with the clay. The methoxy groups in, say, mercaptopropyltrimethoxysilane react with the silanol groups at the kaolin–silica sheet. The pendant mercapto groups react with the double bonds of the elastomer. Huber markets silane-modified clays under the tradenames Nucap (using mercaptosilsilane) and Nulok (using aminosilane). Table 11.2 illustrates their properties [6]. More recently EEC America Inc. has introduced a range of silane-treated clays under the trade

Table 11.2 Nucap and Nulok in SBR 1500

	Compound			
	1	2	3	4
Component				
SBR	100.00	100.00	100.00	100.00
Suprex[a]	175.00	—	—	—
Nucap 100	—	175.00	—	—
Nucap 200	—	—	175.00	—
Nulok 321	—	—	—	175.00
Stearic acid	1.50	1.50	1.50	1.50
Process oil	7.50	7.50	7.50	7.50
Wingstay T (a Goodyear antioxidant)	1.25	1.25	1.25	1.25
MBTS	1.20	1.20	1.20	1.20
DOTG	1.20	1.20	1.20	1.20
Sulfur	3.00	3.00	3.00	3.00
	290.65	290.65	290.65	290.65
Processing				
Scorch MS 3 @ 275°F	12	12	12	9
Mooney viscosity MS 4 @ 275°F	46	42	40	39
Stress—strain Cure 45 min. @ 300°F				
Hardness, points	75	75	75	75
300% Modulus, psi	980	1560	1790	2090
Tensile, psi	2000	2280	2150	2140
Elongation, %	520	450	400	360
Die C tear, RT, ppi	210	240	240	220
Die C tear, 212°F, ppi	110	145	170	125
Compression set B @ 158°F, %	30	18	17	16

[a]Hard clay.

name Polarlink. These clays are spray dried, which means reduced dusting, improved dispersion characteristics and easier bulk handling. They have bulk densities of 700–800 kg/m³, which is almost double that of air-floated or rotary-dried clays. This is a distinct advantage in open mill mixing. The EEC America clays were reviewed by Barbour and Rice [5].

The chemically modified clays do have some attractive properties for the compounder, compared to regular hard clays and some other fillers. For example, they have somewhat lower Mooney viscosity, definitely higher modulus and tensile strength, good retention of properties on heat aging, and low compression set compared with regular clays. These clays are normally considerably less expensive than such other fillers as fine particle silica, and in many cases a partial substitution of these fillers may be made with little if any reduction in performance. Typical uses for these treated clays include inner liners for tires, where they can reduce air permeability, and white sidewalls.

III. SILICA AND SILICATES

Fine particle silica gives the utmost in reinforcement in rubbers of the nonblack fillers; the silicates would probably be in second place. Silica can be made by pyrogenic or precipitation processes. Silicates for rubber compounding are produced only by precipitation.

Silica produced by a thermal method is frequently called fumed silica. The original process was developed by the Degussa Company in Germany in the 1940s and the product marketed, as it still is, under the band name Aerosil. Since that time other producers such as Cabot Corporation have entered the field using variations of the original process. As developed by Degussa, this silica is made by the flame hydrolysis of readily volatile silicon tetrachloride. Silicon tetrachloride, hydrogen, and oxygen are burned in a cooled combustion chamber. The reactions are:

$$2H_2 + O_2 \rightarrow 2H_2O$$
$$2H_2O + SiCl_4 \rightarrow SiO_2 + 4HCl$$

Fumed silicas are only occasionally used in the polymers we are dealing with because of their relatively high cost. Much of this type of silica is compounded with silicone rubbers. There is one exception, however—in certain oil field supplies, the higher price (over precipitated silica and other fillers) is justified. Nitrile rubber is the major polymer used in oil field equipment, where the very severe service conditions include high temperatures and pressures as well as exposure to corrosive fluids, oils, and gases. Products formulated with a high ACN content nitrile using a fumed silica of about 200 m^2/g surface area show higher tensile and tear strengths, and abrasion resistance, and lower water swell, than those using a precipitated silica with about 150 m^2/g surface area. As a second example, ethylene propylene rubbers (EPDM) can be used in blowout preventers, where they are in contact with drilling mud. Again tensile strength, tear strength, and abrasion resistance are greater if fumed silica is used instead of precipitated silica at the same loading.

Precipitated silica is often used if highly reinforced tough white or light-colored compounds are wanted. An example of this application is found in nonmarking solid industrial tires. Although it is hard to imagine the total replacement of carbon black in tires by fine particle silicas, a significant amount has been replaced to obtain certain qualities. Silica is used in off-the-road (OTR) tire treads, for improved tear resistance, hence less chunking of treads and more resistance to cutting. Such tires encounter severe service in the logging and mining industries, for example. In conventional tires up to 20 or 25 parts of the black loading may be replaced by silica to help in lowering rolling resistance, improving traction, and ensuring bonding to the textile and steel cords used in the carcass. The improved bonding is gained by the increased resistance to humid environments.

Precipitated silicas are made by the action of acids on water glass. The acid is usually sulfuric, the water glass an alkaline solution of sodium silicate, normally having a mole ratio of 2.1 to 3.5. The reaction is:

$$n SiO_2 \cdot 1 Na_2O + H_2SO_4 \rightarrow Na_2SO_4 + n - SiO_2 \cdot H_2O$$

The hydrated precipitated silica is filtered out and washed to remove the sodium sulfate. It is then dried and ground. Grinding reduces the size of the aggregates formed.

Properties of precipitated silicas are given in Table 11.3. The properties given cover a wide range of precipitated silicas. Although perhaps 70% of the world's production is used in elastomers, these products also serve in toothpastes, glues, and as a carrier for liquid active ingredients. For satisfying such diverse requirements, a very wide range of silica is made. Those used in rubber, for example, have surface areas at the lower end of the scale—probably no more than $190 m^2/g$.

The moisture that is driven off at 105°C must be carefully controlled in silicas for rubber use. If it is reduced below 3%, adequate dispersion is difficult to obtain. Precipitated silica particles are much more porous than carbon black. This can be shown by comparing BET surface area values with those calculated from electron microscope photographs. In the BET method it is assumed nitrogen gets into the pores; in electron microscope work the surface area is calculated from the apparent diameter of a large number of particles, so porosity is not taken into account. A precipitated silica with a BET of 220 m^2/g may have only 150 m^2/g electron microscopy. Because of this porosity the silica pigments will adsorb accelerators and make them less effective than a nonporous pigment would.

Table 11.3 Properties of Precipitated Silicas

Feature	Amount
Drying loss (105°C), %	4–7
Ignition loss (1000°C), %	8–12
SiO_2 (dry basis), %	83–90
BET surface area, m^2/g	45–700
Particle size, μm (arithmetic mean diameter)	10–100
Refractive index	1.45
Density in rubber	2.00

The silicates used by rubber compounders are largely calcium and aluminum silicates. These are prepared in much the same way as precipitated silicas, but the acid is replaced completely or in part by metal salts like calcium chloride or aluminum sulfate. Process conditions can be varied widely to produce different quality silicates as well as silicas. Variables include concentration of the electrolyte, order and speed of reagent addition, reaction vessel size and shape, and type of agitation. Drying of the silicates can be done in the same manner as the silicas. Rotary tube furnaces, turbo dryers, or spray-drying methods are used. Because of the amount of water involved, drying is a major part of the cost in producing precipitated silicas and silicates.

Before going on, the kinds of hydration present might be noted. As these materials are heated, the first water evolved is that which was condensed in the pores. This water comes off by 105°C. As heating is continued past 200°C, adsorbed water (i.e., water fixed by hydrogen bonds to the silanol group Si—OH) is removed, and finally constitutional water is evolved between 700 and 900°C.

Table 11.4, excerpted from a Degussa technical bulletin [7], gives the main analytical features of aluminum and calcium silicates. There are several producers of silica and silicates for rubber. To name a few: pyrogenic silica is marketed by Cabot Corporation and Degussa; precipitated silicas by Pittsburgh Plate Glass (PPG); SiFrance, and Degussa; and silicates by J. M. Huber Corporation, Durham Raw Materials Ltd., and Akzo Chemie GmbH.

Much of the following discussion of the compounding of silica in rubber applies to silicates as well.

Compounding with precipitated silica is very different from compounding with carbon black. This is because silica has a much more reactive surface than

Table 11.4 Properties of Typical Silicates

Feature	Aluminum silicate: P-820	Calcium silicate: Extrusil
BET surface area, m^2/g	100	35
Average primary particle size, nm	15	25
pH	10.4	10
Sieve residue (45 μm), %	0.1	0.2
Moisture (2 hr at 105°C), %	6	6
Ignition loss (2 hr at 1000°C), %	7	7
SiO_2, %	82	91
CaO, %		6
Al_2O_3, %	9.5	0.2
Na, %	8	2

carbon black. Compared to the silicates (whose use is stationary or declining), fine particle silicas reinforce better but have higher viscosities and cure more slowly. Among the silicas themselves we find some correlation with carbon blacks when surface area is compared with raw and vulcanizate properties. In both cases with lower surface area (larger particle size), viscosity is lower and dynamic properties better, but there is less reinforcement.

The addition of silica to a compound rapidly increases the viscosity, so 50 parts is about the upper limit. Even with 15 parts of softener, for example, 55 parts of silica in SBR 1500 will give a Mooney viscosity (1 + 4″ ML 100°C) of about 90. Before much was known about fine particle silica chemistry, it was difficult to add only 22 parts of Aerosil (Degussa's original silica pigment, which is fluffy, not compressed) to natural rubber. Dusting losses were high, and boardlike stocks produced. Now certain not entirely understood practices greatly help in keeping viscosities low when precipitated silicas are used. Natural rubber softeners of vegetable origin, such as tall oil and hydrogenated rosin, are very effective in compounding. It has been shown that as little as 5 parts of tall oil is equal to 30 parts of naphthenic oil when it comes to reducing viscosity. To keep the viscosity at reasonable levels in SBR compounds, one uses a different route. Petroleum oils or resins can be chosen; however aromatic resins are unique in giving smooth extrusions with better tensile strength and tear resistance, along with better abrasion resistance. But major reductions in viscosity are obtained only with additives that de-agglomerate the silica. Whereas there is an affinity between carbon black and nonpolar rubbers, silica is hydrophilic and incompatible. Silica-to-silica attraction is high, with the result that large aggregates are formed, impeding flow, and the mix becomes very stiff. The additives chosen for this situation can be soluble zinc compounds (like zinc octoate) or materials such as hexamethyltetramine (hexa or HMT). Magnesium oxide acts similarly but produces a severe loss in reinforcement. In using these viscosity reducers, dosage might be 3 phr for zinc octoate, 0.5 for zinc methacrylate and HMT. On remilling a typical silica-reinforced SBR compound without these materials, the viscosity was 81, but the same stock with the additives above at the dosage noted showed viscosities of 57, 52, and 68, respectively.

Another unique feature of silica compounding is the good activating effect of glycols, particularly polyethylene glycol (PEG). Usually PEG is used at 2 parts per 100 of rubber. The activation effect shows up in significant reductions in cure time, plus higher tensile strength; compression set and heat buildup are lower. Triethanolamine also is used for its activating effect.

A wide variety of curing systems is used with silica, depending on the polymers used and the properties wanted. The subject is too broad to be handled adequately here. A major supplier of precipitated silicas, PPG Industries, has contributed many papers on the use of this material. A valuable aid to compounders of precipitated silica is a report by N. L. Hewitt of that company [8].

Mention has been made of the good effects that can be achieved by the use of coupling agents with clays. These materials are even more effective in silica compounding. Viscosity is reduced, modulus and tensile strength increased dramatically, and the abrasion resistance increased significantly. Two coupling agents are most frequently used, mercaptopropyltrimethoxysilane and bis(3-(tri-ethoxisilyl)propyl)tetrasulfane. A typical commercial example of the first is A189, made by Union Carbide; the second is represented by Degussa's reinforc-ing agent Si 69. The manufacturer of Si 69 explains the reinforcing action by noting that the chemical reaction occurs via the filler-reactive ethoxisilyl groups with the silanol groups of the silica surface, which form stable siloxane bonds. Also, chemical reactions between the tetrasulfane group and the elastomer during the vulcanizing process link the silica filler with the elastomer matrix.

There is ample evidence of how well these agents work. Hewitt [9] describes how a 60-part silica loaded S-1502 stock was tested with and without 1.2 parts of mercaptosilane. For the stock using the coupling agent, viscosity dropped from 103 to 77. Tensile strength increased from 2700 psi to 4200 psi and road wear index increased from 79 to 114. In another test it was reported that a 130 m^2/g silica using Degussa's Si 69 in a natural rubber tread compound gave the same tread wear as a 135 m^2/g carbon black compound [10]. The high price of the agents may be one reason silica compounds with coupling agents are not used more extensively in tires.

Depending on the polymers used and compounding techniques, a wide vari-ation in properties can be obtained using silica and silicate fillers. Tensile strengths can go up to 4200 psi, elongations up to 850%; hardness can be varied from 40 to 80 Shore A. Compression set (ASTM method B) of around 70 after 3 days at 100°C may be high but can be considerably reduced by going to an efficient curing system. In noncoupling agent stocks polyethylene glycols such as Carbowax 4000 are often used to improve processing and tensile properties.

IV. CALCIUM CARBONATE

Calcium carbonate is a widely used white filler. It occurs naturally as calcite and aragonite, and any rock containing over 50% by weight $CaCO_3$ is called lime-stone. The natural product is quarried or mined, then crushed and ground. The ground material, also called whiting, is used in rubber. Frequently the ground material is used as a base to make precipitated calcium carbonate and sold to the rubber industry under various trade names. These brands have finer particles than ground limestone and, to ease dispersion, the particles may be coated with rosin or stearic acid. This surface coating may run 2–3% of the pigment weight.

Densities of calcium carbonates vary widely depending in part on whether the product is mainly calcite or aragonite and whether it is ground or precipitated. Although densities may run 2.5–2.9, ground limestones usually run 2.65–2.70, precipitated 2.50–2.65. The precipitated grades, not surface-treated, usually assay over 98.5% $CaCO_3$, are whiter and have surface areas of 10–30 m^2/g measured by the BET method. The arithmetic mean particle size of precipitated carbonates, can vary rather widely from 30 to 350 nm.

Like clays, calcium carbonates are used at higher loadings than carbon blacks: 50–150 parts is not uncommon. They are of little reinforcing value, but even at these loadings they process well and are economical; their compounds are still quite rubbery. Product lines using these pigments include hot water bottles, footwear, and extrusions. It should be noted that calcium carbonate is not resistant to acids, so its pigments should not be used in such items as acid hose tubes.

A more reinforcing calcium carbonate than most is a product called Fortimax made by ICI Mond Division, Cheshire, England. This precipitated calcium carbonate has been treated with 3% carboxylated polybutadiene. The process changes the carbonate to a hydrophobic material that disperses readily. Polymer interaction is much improved because the carboxyl groups are bonded to the calcium carbonate and the polybutadiene segments are vulcanized along with the rubber matrix. Tested in SBR 1502 at a 40-volume loading against a stearic acid coated calcium carbonate, Fortimax showed much higher viscosity, modulus, hardness, and abrasion resistance [1].

Before leaving calcium carbonates, a unique form should be mentioned. Oyster shells are largely calcium carbonate, and large reefs of these shells occur in certain coastal areas. When coarsely ground, the shells are used as a food supplement in the poultry industry; when more finely ground they can be used as a compound ingredient. This product is unique in that the particles are coated by nature with a polymolecular proteinaceous coating, which assists dispersion tremendously so that physical properties are improved, and it gives especially smooth extrusions. It can also be used in higher loadings rather than coated precipitated whitings or clay without excessive stiffening. Developed in the United States, the product was initially marketed as Lamina but is now called Pilot SF4. If the proteinaceous matter is removed by enzyme action, the product loses its advantage over other whitings.

REFERENCES

1. Dannenberg, E. M., Filler Choices in the Rubber Industry, *Rubber Chem. Technol.*, 55:862 (1982).
2. Byers, J. T., Non-Black Fillers in a Rubber Compound, Philadelphia Rubber Group Meeting, Sept. 19 (1991).
3. *Kaolin Clays and Their Industrial Uses*, J. M. Huber, Macon, Ga., 1949, pp. 66, 67.
4. *Kaolin Clays and Their Industrial Uses*, J. M. Huber, Macon, Ga., 1949, pp. 67, 68.

5. Barbour, A. L., and Rice, A., New Clays Improve Compound Properties, R. & P. N: in Rubber and Plastic News, Aug. 24, 1987, p. 48.

6. Florea, T. G., *Nucap and Nuok as Reinforcements in Elastomers*, Table IV, technical bulletin, J. M. Huber Macon, Ga.

7. *Precipitated Silicas and Silicates: Manufacture, Properties and Applications*, Degussa bulletin, Teterboro, N.J July 1984.

8. Hewitt, N. L., Compounding S lica in Elastomers, Rubber Compounding Report, PPG Industries, Inc, Pittsburgh. Pa.

9. Hewitt, N. L., Processing Technology of Silica Reinforced SBR, *Elastomerics*, March 1981.

10. Wolff, S., *Effects of Bis-(3-triethoxisilylpropyl)-tetra-sulfide-modified silicas in NR (Part 1)* presented at the ACS Rubber Division meeting, Washington, DC, September 1979.

12

Processing Aids

I. PEPTIZERS

Processing aids can be defined in various ways. No definite line can be drawn between process aids, for example, and the plasticizers and softeners discussed in the next chapter. Nevertheless process aids will be considered here to be the materials included in a mix primarily to reduce the time and energy required in breaking down the polymer. Frequently, however, at the same time they help with dispersion of dry materials, give smoother stocks, improve extrusion rates, and in some instances increase the homogeneity of blended rubbers. They are generally added in small quantities, usually less than 5 parts per hundred rubber (phr), and the effect on vulcanizate properties is minimal.

RSS #2, a common grade of natural rubber, might be received at the plant with a Mooney viscosity (ML 1 + 4 min @ 100°C) of 85–105. To attempt to mix carbon black into that rubber without breakdown would be like trying to drive ball bearings into a block of hardwood with a hammer. By using a mill, Banbury, or other plasticating device, the rubber instead would be masticated to perhaps 60–75 Mooney, at which point incorporation of carbon black would be more feasible. The process requires considerable energy, and any material that shortens the time or energy cost is certainly a process aid. If this breakdown is accelerated by a process aid that works more in a chemical way than a physical way, the agent is called a peptizer.

A word of caution might be appropriate here. Products in this field are changing comparatively fast. Materials used for decades are being withdrawn from the market, suspected of having unacceptable toxicity. New products with unique capabilities such as homogenizing agents to improve the mixing of blends of polymers have been introduced. Suppliers' literature must be consulted to keep abreast of developments.

A study has been made [1] comparing three brands of chemical peptizers in their ability to break down RRS #1 in Banbury mastication compared to mastication without them. The three peptizers were Renacit VII from Mobay Corporation, Pepton 44 from American Cyanamid, and Struktol A82 from the Struktol Company of America. These have been identified (in the same order) as pentachlorothiophenol with activator and dispersing additives, activated dithio-bisbenzaniline on an inert carrier, and a blend of high molecular weight acid ester and active chemical peptizer.

The Mooney viscosity, originally 83, was almost linearly brought down by mastication alone to 62 after 10 min of processing. For many purposes a Mooney viscosity of 60 ± 5 would be appropriate breakdown before adding other ingredients.

The softening action of the peptizers was readily apparent. All three peptizers brought the rubber to the 60 ± 5 range within 2–5 min of mixing. This was accomplished at levels as low as 0.1 phr for the Renacit VII- and Pepton 44-treated rubbers, 0.4 phr for Struktol A82. Selection of a peptizer would not of course be made solely on effectiveness per pound in reducing viscosity; cost and other technical properties would be considered as well.

The use of peptizers exacted a price in terms of stress–strain results on the vulcanizates. The rubbers were mixed in a standard natural rubber tread formula using the batches with the highest amount of peptizer (viz., 0.50 part Renacit VII and Pepton 44, 2.0 parts Struktol A82). Even at these high levels, for batches mixed 6 min the tensile drop was 130–630 psi, the 300% modulus loss up to 300 psi. With the control (no peptizer) having a tensile strength of 3980 psi and a 300% modulus of 2000 psi, these deficiencies, while significant, cannot be considered large. Despite hot air oven aging for 5 days at 100°C, the peptized rubber vulcanizates did not show significant deterioration compared to the control.

Although it includes at least one and possibly two peptizers not marketed or used in the United States, a publication of the Malaysian Rubber Producers Research Association [2] gives more information. Four peptizers were used, Pepton 22 (di-*o*-benzamidophenyldisulfide), Pepton 65 (zinc 2-benzamidothiophenate), Renacit IV (zinc salt of pentachlorothiophenol), and Renacit VII (pentachlorothiophenol with activating and dispersing additives). The first two are products of American Cyanamid, the other two of Bayer. These were mixed in the laboratory in an internal mixer (BR Banbury) with a technically specified rubber, SMR 5, at peptizer concentrations of 0.05, 0.25, and 0.50 phr. It is understood that extrapolation of such results to full-sized factory mixers may be difficult.

The MRPRA found no linear relation between decrease in viscosity and the amount of peptizer used. The greatest reduction occurs with the first addition of peptizer; doubling the concentration doesn't halve breakdown time. As might be expected, increasing the temperature hastens breakdown. As an average, the four peptizers at 100°C took about 7.4 min to reduce the rubber to a 50–55 Mooney viscosity; at 130°C the average was about 4.3 min and at 160°C 1.2 min. The ability of the peptizer to reduce viscosity with increasing temperature changes with the brand. For example, Pepton 22 was the slowest at 100°C but the fastest of the four at 130°C in reaching the 50–55 level.

Peptizer use is mainly confined to natural rubber compounds. Aromatic mercaptans, formerly the main agents used, have been replaced by other chemicals largely for safety considerations in the workplace. Chelate complexes of iron, cobalt, and manganese accelerate the breakdown of natural and some types of synthetic rubber by thermal oxidation. Peptizer dosages of 0.1–0.5 phr are usually sufficient. The use of peptizers for the mastication of rubber is well outlined by Fries and Pandit [3].

II. SEMIPEPTIZERS

Although the chemical peptizers described above are most efficient in reducing viscosity, certain other materials might be called semipeptizers because they do assist in the mechanical breakdown of natural and synthetic rubbers. These are generally mixtures of sulfonated organic compounds and mineral oils with densities in the range 0.83–1.23. Usually they are liquids, but they can come in powder or friable cake form. Typical trade names are Reogen, marketed by R. T. Vanderbilt Company, and Peptizer 965, sold by the C. P. Hall Company. The amount used is considerably higher than with chemical peptizers, probably 1–3 pph, occasionally 5 pph in softer stocks like sponge. Besides the mild peptizing action, these materials often expedite mill and mold release.

III. PROCESSING/DISPERSING AGENTS

Within the scope of this chapter, materials termed "processing and dispersing agents" by their manufacturers are the most numerous and widely used. Variable in composition, they may be subdivided into low molecular weight resins, fatty acids, fatty acid esters, and fatty acid metal soaps.

An example of a low molecular weight resin used as a processing and dispersing agent is Kenflex A, marketed by the Kenrich Petrochemicals Inc. Kenflex A is produced by condensing selected aromatic hydrocarbons with formaldehyde, resulting in resin molecules that have two or more methylene bridges between fused aromatic rings. Because the high polarity of the fused aromatic rings exerts an activating influence on the hydrogen atoms on the adjacent methylene groups, the formation of hydrogen bonds is facilitated, leading to a relatively high degree of solvating and hydrogen bonding with most high molecular weight polymers. Since the oligomeric resin is tightly bound to the surrounding high polymer matrix, it is difficult to remove by extraction or volatilization. Kenflex A is a light amber solid with a density of 1.08; it has a melting range of 58-74°C. The calculated molecular weight is 662.

Some rather interesting results are reported in a study of the effect of Kenflex A on plasticity and scorch on a conventional neoprene W compound [4]. The mix contained 20 parts of FEF black and 90 parts of hard clay. As the resin content was increased to 40 parts on the polymer, the Mooney viscosity dropped from 70 to 41, yet the original stress–strain properties were not changed in a major way. Tensile strength at 2050 psi was equivalent to that at 2010 with no resin; hardness at 40 parts resin was 79 Shore A units, only 3 units below the 82 shown by the control. Obviously the processibility of the neoprene W stock was increased.

A unique property manifested itself in a related study, again with neoprene W, where the resin was compared with a light naphthenic oil and butyl oleate. Neoprene can crystallize fairly rapidly at low temperatures, and this is a disadvantage in some applications. The rate of crystallization can be measured at −10°C by noting the hours required for a 20-point increase in durometer hardness. This required 264 hr for the resin-modified stock, 40 hr for the naphthenic oil stock, and 11 hr for the butyl oleate mix. All process aids were used at the same dosage (15 phr). It is theorized that the resin molecules are so large that they can physically prevent polymer orientation, thus preventing crystallinity.

There are several manufacturers of low molecular weight resins that can be considered to be processing and dispersing aids. For example, low molecular weight resin blends are marketed by the Struktol Company of America as homogenizing agents. These are sold as 40 MS, 40 MS flakes, and 60 NS flakes, where MS stands for minimal staining and NS for nonstaining. The composition of these resins is not disclosed. Each has a melting point range of 10°F with the lowest melting point range 140–150°F, the highest 200–210°F, the flakes having 40–60°F higher melting point ranges. The 60 grade is more polar than the 40.

The products are resin blends and have constituents that are compatible with the aliphatic, naphthenic, or aromatic parts of elastomers in an elastomer blend. It is for this reason that they can improve blend homogeneity. This can also be done if a mix of plasticizers, compatible with the elastomers of the blend, is added to the mix. However, this may lead to migration and bloom of the plasticizer. Klingensmith [5] reports two examples of the utility of using a homogenizing agent. A conventional passenger tire tread compound using 70 parts SBR and 30 parts BR was experiencing shrinkage and die swell during processing. By adding 4 parts of a homogenizing agent (Struktol 40 MS) to the mix, shrinkage was halved with no adverse effects noted. Again in an NR/BR sand blast hose tube stock, 5 parts of the homogenizing resin and 2 parts of a fatty acid ester were added. As a result, the mixing cycle was reduced from 4.5 to 2.5 min, yet both tear and abrasion resistance were improved.

Finally we come to the processing aids that are centered around fatty acids. Fatty acids and their derivatives serve as lubricants between the polymer chains and allow easier flow of the compound in shear. This is possible because these materials have a restricted solubility in rubber compounds. In such operations as injection molding, shear ratios are very high, and fast-flowing compounds are necessary for proper mold fill in the short time available with current production schedules.

Stearic acid as a processing aid has a low solubility in many polymers and is not usually used at over 3 phr. It helps prevent compound from sticking to mill and calendar rolls. Stearic acid is used in somewhat larger quantities, say about 5 phr, in many varieties of interconnecting cell sponge rubber. Here in addition to plasticizing, the compound, it helps in the liberation of carbon dioxide from such blowing agents as sodium bicarbonate. Both ends of the stearic acid molecule help processing. The paraffinic end provides the lubricating effect, and the acid group absorbs on the surface of pigments—this pigment wetting ability is the second advantage. Although stearic acid is by far the most common fatty acid used, others, like oleic acid or palm oil containing fatty acid, may be used for special purposes.

As one goes from fatty acids to derivatives, the choices multiply, and clear descriptions of their compositions and how they function become less frequent. These derivatives fall into two broad classes, fatty acid soaps and fatty acid esters. The fatty acid soaps are prepared by the classic saponification reaction shown in Figure 12.1, where MOH is the metal hydroxide with M usually representing Na, K, Mg, Ca, or Zn. Fatty acid esters are made by the interaction of an alcohol with the fatty acid:

$$RCOOH + R'OH \rightarrow RCOOR' + H_2O$$

Probably the most common fatty acid soap used in rubber, zinc stearate most often serves as a surface coating for raw stock but can be used as an internal release agent. Zinc salts of unsaturated fatty acids, such as zinc oleate, are physical peptizers for natural rubber. Here they stabilize thermal and oxidative

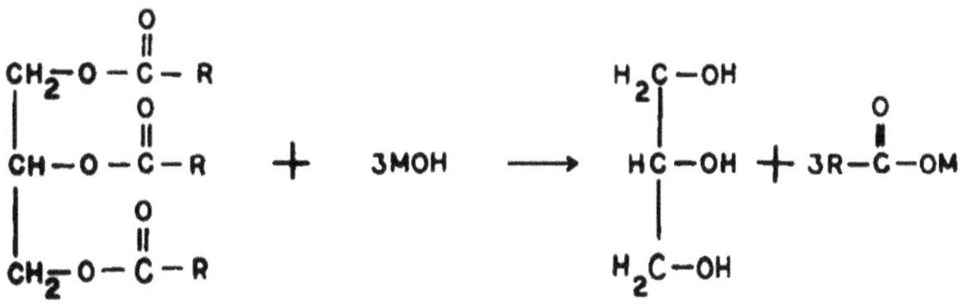

Figure 12.1 Classic saponification reaction.

chain scission. Fatty aid esters are used if it is important to avoid bloom. They are more compatible in rubber compounds than the acids or soaps.

It is difficult to pigeon-hole some of the specialty chemicals used in rubber compounding. A case in point is a new product of the Struktol Company of America called Struktol Activator 73, marketed as a multifunctional cure activator and described in a paper by Wasko [6]. Inasmuch as the product gives improved processing, it is included in this chapter.

Activator 73 (A73) is a zinc soap, a composition of selected aliphatic and aromatic zinc salts of carboxylic acids. Mooney viscosity in natural rubber dropped from 68 to 59 after the addition of 3 parts of A73. The natural rubber compound had 60 parts of N220 back, 5 parts of oil, and a curative system of 1.00 part of TBBS and 1.50 parts of sulfur. Perhaps even more noteworthy was the reduction in energy consumption in mixing an SMR 5 rubber batch, where energy consumption was only 72% of the control batch. In a simulated tread stock of solution SBR, addition of 3 parts of A73 made processing easier. The treated stock had a significantly greater flow rate in a capillary rheometer, with a flow of 155 versus 109 (ML/sec $\times$ 10^{-4}), viscosity (ML 1 + 4 min @ 212°F) was 68.4 against 75.2 for the control. Particularly impressive was the cohesive tack from 0.29 lb to failure (1 in. ~ plied area) to 2 lb when A73 was used. All these properties are processing advantages. There were vulcanizate improvements, such as increased reversion resistance and faster curing, which are elaborated on in the paper cited [6].

IV. COMPOUNDING WITH PROCESS AIDS

Some idea of the range and complexity of processing aids was shown earlier by Klingensmith [5]. A summary list included 21 brands with prices ranging from 64.5 cents a pound to $16 a pound. Compositions included numerous blends, and besides the fatty acids and their derivatives noted earlier such diverse materials as high density polyethylene, sulfonic acid ester, and sodium alkyl sulfate were present.

Unless it is a case of using peptized natural rubber, compounders rarely include a process aid when formulating a new compound. This is understandable because the primary aim is to get the desired vulcanizate properties. Processibility is of course necessary but is examined second.

If a compound gives appropriate vulcanizate properties but doesn't run right (e.g., extrusion rates are too slow) and changing the processing conditions fails to rectify matters, a processing aid should be considered. Here laboratory trials may be of limited value. For example, Mooney measurements, while helpful in many ways, do not use the shearing ratios employed by factory equipment. It is a considerable task to extrapolate laboratory tubing results from, say, 3/16-in. die extrusions to the shrinkage obtained in a 4-in. cylinder diameter tuber. Under these circumstances the kind of processing difficulty should be determined as closely as possible and then the help of suppliers solicited to determine the additive to be used for a factory trial.

REFERENCES

1. Goodyear Plantations, Natural Rubber Compounding Studies, *Natural Rubber Banbury Peptization*, Goodyear Tire and Rubber Co., Akron, OH.
2. Technical Information Sheet 082, MRPRA, Brickendanbury, England, 1979.
3. Fries, H., and Pandit, R. R., *Mastication of Rubber*, in Rubber Chemistry and Technology, *1982*, Vol. 55, 309.
4. Kenrich Petrochemicals Inc., *Neoprene Processed with the aid of KENFLEX A, KENFLEX N, KENMIX Red Lead, KENLASTIC Red Lead, KEN-MAG*, technical bulletin, Bayonne, N.J.
5. Klingensmith, W. H., *The Effects of Processing Aids in Rubber Compounds*, Chicago ACS Rubber Division Meeting, October 1982.
6. Wasko, J. J., Jr., *Struktol Activator 73: A Multifunctional Cure Activator and A Review of Zinc Soaps*. Paper given to the Los Angeles Rubber Group, Nov. 5, 1991.

13

Plasticizers and Softeners

I. INTRODUCTION

As mentioned in Chapter 12, it is impossible to draw a distinct line between process aids and plasticizers and softeners. The ASTM definition for softener is "a compounding material used in small proportions to soften a vulcanizate or facilitate processing or incorporation of filler"; plasticizer is "a compounding material used to enhance the deformability of a polymeric compound." These definitions describe the materials reviewed in this chapter.

Two points might be noted. Certain types of compounds, such as high black, high oil EPDM mixes, may well have oil loadings exceeding half the weight of the rubber. The oil is undoubtedly a softener, yet this amount would hardly be called a small proportion. Note also the different shade of meaning for a plasticizer compared to a softener. An ester like dioctyl sebacate might be added to a neoprene W composition: undoubtedly it has a softening action, but primarily it is used to give greater flexibility at low temperatures than would otherwise be possible.

A difficulty in assessing these ingredients is the sheer number of them. During World War II an exhaustive study [1] was made of softeners so the then-new synthetic rubber, styrene-butadiene rubber, could be used most effectively. That study used over 600 different materials. Undoubtedly some have been taken off the market but probably not nearly as many as have been added. Many of these products are proprietary, ill-defined in composition, and coming from a multitude of sources. Selecting the ideal one for a compound from such an array is not easy.

As with so many facets of compounding, it is not known exactly how softeners and plasticizers achieve their results. Some have a solvent like effect on rubber. Perhaps the simplest manifestation of this is benzene and natural rubber. Here the benzene is imbibed so much by the polymer chains that under certain conditions (e.g., temperature, agitation) the rubber piece loses its form and you have a solution of the rubber in benzene. Benzene is not used as a softener, but this solvating effect is present with such materials as pine oils and aromatic tars. It is obvious that many softeners and plasticizers operate differently—coumarone indene resins, for example. There are some general rules that indicate the strength of the solvent action on polymers. Aromatic solvents are more effective solvents of aromatic polymers than aliphatic, and aliphatic solvents are more effective with aliphatic polymers than aromatics. Such fluid effects on polymers are reviewed by Files [2].

Apart from solvating causing softness, several theories have been proposed. In an overview, Stephens [3] recognized three as prominent. These are the lubricity theory, the gel theory, and the free-volume theory. The first is perhaps the most easy to visualize. A polymer might be construed as a nonwoven fabric like felt. The individual fibers are like the polymer chains. Although randomly oriented, they are so entwined that moving them is difficult. If a plasticizer is introduced between the chains, they become physically separated and can slide past one another more easily; consequently the whole mass is more deformable.

In the gel theory the stiffness of the polymer chain is due to a three-dimensional structure or gel-like network, active sites along the chains serving as hooks as it were between chains. Adding a plasticizer serves to mask these points of attraction and there is a swelling of the gel, so deforming is made easier.

The free-volume theory holds that the purpose of the plasticizer is to lower the glass transition temperature T_g of the polymer. Doing this increases the free volume, which is a function of temperature and structure. There is some question as to whether these theories might not be more applicable to plastic compounding than rubber. Readers desiring further information are advised to consult the references in the overview [3].

As can be imagined with the wide variety of materials claimed to be softeners and plasticizers, there is no shortage of classification schemes. Most use as a base the source from which material is made, and this is done here. In terms of weight, more than 99% of these agents fall in the following classes:

1. Petroleum
2. Pine tar and pine products
3. Synthetic resins
4. Esters
5. Natural fats and oils

II. PETROLEUM-BASED SOFTENERS

The softeners most widely used by compounders are undoubtedly petroleum oils. There are several reasons for this: these oils are versatile, effective, tightly controlled for quality, economical in price and, as liquids, easy to use. Recently there has been concern about the medical acceptability of some rubber process oils. For example, the U.S. Occupational Safety and Health Administration (OSHA) adopted the report of the International Agency for Research on Cancer (IARC), which classifies untreated naphthenic oils as carcinogenic. An untreated naphthenic oil is one that has not seen sufficient processing such as severe hydrogenation or extraction. Accordingly, oils that have more than 0.1% of an untreated naphthenic oil must be accompanied by a Material Safety Data Sheet (MSDS) stating that it contains untreated naphthenic oil, which is considered carcinogenic. Similarly, an oil containing more than 0.1% polynuclear aromatics (PNA) will need a statement on the MSDS that it contains polynuclear aromatics, which are considered carcinogenic. There are a number of applications for untreated naphthenic oils. Further restrictions may be introduced, so compounders should check with their suppliers to be kept up to date.

Fundamentally, rubber processing oils are high boiling fractions obtained in refining after gasoline, fuel oil, and other low boilers are removed by distillation. Rubber process oils cannot be classified by chemical composition directly. Few of the myriad chemicals in lubricating oils or rubber process oils have been identified. Some idea of their composition can be obtained by correlation with the physical properties of pure high molecular weight compounds.

Oils are made up largely of ring structures, with an oil molecule probably containing saturated rings (naphthenes), unsaturated rings (aromatics), and rings with saturated side chains (paraffins). If by indirect analytical means the major portion of the rings appears to be unsaturated, it is classified as an aromatic oil; if saturated rings prevail, it is called naphthenic; but if the naphthenic rings are fewer and a large number of saturated side chains exist, the oil is classified as paraffinic. Although the name paraffinic is used, the main source of paraffin— paraffin wax—has been removed.

Since the composition of an oil has an effect on the properties of the rubber compound in which it is included, various identification methods have been developed. Two major analytical methods are used: molecular and carbon-type analysis. ASTM designation D2226 classifies petroleum extender oils used in extending and processing styrene and butadiene rubbers. It separates oils into four types: 101, 102, 103, 104. Types 101 and 102 are considered to be aromatic types, 103 naphthenic, and 104 paraffinic. D2226 classifies oils by using a molecular-type analytical method (D2007), where the oil is adsorbed in a chromatographic column containing clay and activated silica gel. This method analyzes the oil for asphaltene, saturates (naphthenes and paraffins), polar compounds, and aromatics. Material insoluble in pentane is considered to be asphaltene, although it may include oxidation products such as insoluble resinous bitumens.

This method does have limitations in rating an oil for rubber compounding purposes. The amounts of aromaticity and naphthenicity are not clearly distinguished (e.g., a benzene ring with an attached naphthenic group would be rated as all aromatic). For such reasons carbon-type analysis is also used. A method developed in the laboratories of the Sun Refining and Marketing Company depends on the correlation between the viscosity–gravity constant (VGC) and the refractivity intercept. The viscosity--gravity constant is calculated from the specific gravity of the oil at 60°F, the Saybolt viscosity at 100°F. The VGC is another way of measuring the aromaticity of an oil and is calculated as follows:

$$VGC = \frac{10G - 1.0752 \log (V - 38)}{10 - \log(V - 38)}$$

where G is specific gravity at 60°F and V is Saybolt viscosity at 100°F.

The refractivity intercept is a ring index calculated using the refractive index at 20°C for the D line of sodium and the density of the oil at 20°C. The refractivity intercept is easily calculated as follows:

$$\text{refractivity intercept} = n_D^{20} - 0.5d_4^{20}$$

where n_D^{20} is the refractive index at 20°C for the D line of sodium and d_4^{20} is density at 20°C.

After these constants have been determined, they are plotted on a unique triangular graph [4]; from their point of intersection, the percentage of carbon that is naphthenic, aromatic, or in paraffin chains can be read off.

Several physical properties of rubber processing oils are of interest to the compounder. These include such measurements as specific gravity, volatility, molecular weight, and color. The importance of specific gravity has already been illustrated by its use in determining the VGC. One supplier's line has a specific gravity range from 0.847 to 0.998, a significant spread in determining the density of high oil loaded batches. Volatility for the same line, measured at 225°F, varied from as little as 0.26% to a high of 19.3%. Obviously one would not choose the latter oil for a product exposed to high temperature in service for extended periods of time. Molecular weight range effects and color are reviewed later.

As might be expected the compatibility and processing help given rubbers depends on the composition of the oil and the type of rubber involved. Certain guidelines have been established through experience. For all rubbers, as the oil content is increased one can expect the viscosity of the mix to decrease and time for incorporation of the oil to expand. In the vulcanizates, higher loadings of oil result in lower tensile strengths and modulus, higher elongations, and lower hardness. High aromatic oils are most often used with SBR, BR, and CR rubbers, since these polymers are more resistant to naphthenic and paraffinic oils. Naphthenic oils are commonly used with EPDM. Paraffinic oils are commonly selected for natural rubber and butyl compositions.

A. Properties and Compounding

Table 13.1 gives the range of several oil properties that might be expected from oil suppliers. Within each type, a manufacturer may supply several grades. The table can be used in selecting first a particular type of oil for the compound in mind and then, within that type, the manufacturer's grade that seems most appropriate for the service conditions. There are two main considerations in choosing a petroleum oil as a softener or plasticizer: (1) its effect on processing behavior and (2) its effect on vulcanizate properties, as discussed in the sections that follow.

1. Viscosity

The viscosity of an oil is of concern to the compounder with respect to handling and the properties it imparts. With highly viscous oils handling becomes difficult for accurate and quick addition to the batch. This can be alleviated if oil-heating facilities are provided or if the oil can be dispersed on a carrier and then introduced.

If viscosity is high, molecular weight is generally high and compatibility with the polymer is less. This means that more mixing time is required for full

Table 13.1 Range of Property Values for Petroleum Oils

Property	ASTM method	Naphthenic	Aromatic	Paraffinic
Saybolt viscosity				
SUS, 100°F	D2161	108–3600	100–9800	65–2540
SUS, 210°F	D2161	38.2–113	37.3–300	35.5–159
Flash point COC, °F	D92	340–503	340–575	350–579
Pour point, °F	D97	−40 to +20	−35 to +95	0 to +10
Aniline point, °F	D611	154–229	89–165	195–267
Clay gel analysis, wt %	D2007			
asphaltenes		0	0	0
aromatics		20.2–48.3	57.9–74.3	10.0–23.4
polar compounds		0.1–4.5	3.5–15.9	0.1–3.8
saturates		47.9–79.7	15.5–31.6	74.1–89.9
Carbon type analysis, %	D2140			
$C_{aromatic}$		6–21	29–41	1–5
$C_{naphthenic}$		29–43	10–37	22–37
$C_{paraffinic}$		41–62	33–53	59–73
Pounds/gal. at 60°F	—	7.39–7.89	7.80–8.38	7.10–7.45

The petroleum oils columns (Naphthenic, Aromatic, Paraffinic) fall under the heading "Petroleum oils".

dispersion of ingredients. Apart from this processing concern, generalizations can be made about the effect of viscosity on vulcanizates. Although the rate of change of viscosity with temperature is the guiding principle, oils with high viscosities usually cause less flexibility at lower temperatures. Tensile strength and hardness tend to be higher with higher viscosity oils, resilience less. These characteristics may not be general for all polymers. One paper [5] reports no change in vulcanizate properties when a series of naphthenic oils with a wide variation in viscosity was evaluated in EPDM. If low temperature flexibility is important, it should be remembered that as the temperature is lowered, aromatic oils increase in viscosity faster than paraffinic ones.

2. Specific Gravity

Paraffinic oils have the lowest specific gravity, followed by the naphthenic and then the aromatics. Note that the highest density oil is about 20% higher than the lowest. These differences are great enough to force careful consideration of what compound density to expect when high amounts of oil are used in high oil, high black compounds. Also differences like these may necessitate loading adjustments if equal volumes of oil are wanted in a study.

3. Pour Point

Pour point values of petroleum oils are most often considered in relation to an oil's ability to lubricate well under low temperature conditions. But they are also another guide to the compounder in selecting an oil to give low temperature flexibility. The lower the pour point, the more suitable the oil. Naphthenic oils are best at imparting low temperature flexibility, paraffins next, and aromatics last.

4. Aniline Point

The aniline point of an oil, the temperature at which aniline and the oil becomes miscible, is determined in accordance with ASTM method D611. It is inversely related to VGC value. Low aniline points indicate high aromaticity, and oils with low aniline points are thought to be more compatible with polymers like SBR and neoprene or nitrile. Oils are so complex that it is very difficult to select one characteristic like this as indicative of the processibility of the oil with a polymer.

5. Volatility

The volatility of an oil is usually measured by the loss in weight after 22 hr at 225°F. The volatile loss for rubber oils rarely exceeds 10% and, as noted before, decreases with increasing viscosity However, this test is conducted at 225°F, since rubber compounds are mixed, cured, and sometimes tested at higher temperatures, volatile losses are of concern. For example, temperatures in the Banbury might reach 300–350°F. Dimeler [6] studied these situations and he found a maximum of 4% loss of oil in an SBR stock in circulating air after 6 min at 325°F. This occurred with the most volatile oil of seven tested. Most had losses of 2% or less. This simulates the losses that might occur in a Banbury. Losses during press cures are insignificant—less than 1% at normal curing times. Oil losses can be high, however, if vulcanizates must meet severe thermal aging requirements. With the more volatile oils, 20–30% of the oil may be evaporated from SBR cured sheets in circulating hot air at 325°F in 45 min.

6. Color and Staining

Aromatic oils are usually dark in color and so are rarely used in white stocks. In oils polar compounds are heterocyclic compounds containing nitrogen and sulfur. Polar compounds tend to concentrate in the aromatic oils during refining because of similarity to structure. However, these polar compounds reduce the oxidation stability of an oil and therefore cause it to discolor during the action of UV light. As a first approximation, then, the lower the aromatics, the lower the staining.

B. Oil Composition

Before noting more specifically the effects of oil composition, it might be useful to describe what happens when oils and rubbers are combined. With increasing oil we can expect softer, more easily mixed stocks. However, if high volumes of oil are added at the start of the mix, dispersion of carbon blacks or other reinforcing fillers will be poor and vulcanizate properties mediocre. To avoid problems associated with poor dispersion, oils are added part way through the mix, often in increments. An exception occurs in so-called upside-down mixing: that is, mixing in an internal mixer, where the fillers and oils are added first, then the rubber—opposite to conventional mixing techniques. An oil is increased in the vulcanizate, tensile strength and modulus are lowered, elongation increased, and hardness decreased.

It should be realized, too, that there is a definite limit on the amount of oil that can be compatible with an elastomer. If this limit is exceeded, the oil will sweat out, probably making the vulcanizate unfit for use. There are no precise guidelines here. As a matter of interest, however, neoprene will not tolerate much more than 5 parts of a paraffinic oil and 25 parts of a naphthenic oil, but more than 100 parts of a highly aromatic oil.

One compounding expedient might be noted here. Occasionally specifications limit the volume swell of a rubber piece in oil. If that oil can be used in the compound mix, there will be reduced swell when the cured piece is tested. If the amount of oil added to the mix is the equivalent to the volume percent of oil at equilibrium swelling for the vulcanizate, there should be no volume change when the product is tested. To meet volume swell specifications by adding a softener extractable by the test medium is shortsighted if the extracted material will contaminate or spoil the utility of the liquid with which it is in contact.

The oil component that has the most effect on rubber is the aromatic content. Most studies on oil composition and rubber properties have dealt with SBR largely, because of the huge volume of oil-extended SBR marketed. The remarks that follow refer to SBR compounding unless otherwise noted.

The rate at which a coming together of the ingredients is accomplished, what can be called the oil takeup time, increases with increasing aromaticity in an oil. Also, the dispersion of carbon black is better, naphthenic oils are second, and paraffinic oils last. At least in the early stages, highly aromatic oils give faster rates of cure, and this result has been attributed to the heterocyclic sulfur and nitrogen compounds they contain. These same compounds may cause poor aging of oil-extended polymers on long storage. Tensile strength and tear increase with aromatic content; this may well be due to better dispersion of the carbon black. Rebound resilience is inferior to that of paraffin oils, but crack resistance is better. Again the hetereocyclic sulfur and nitrogen compounds are subject to stain and discoloring problems as aromaticity increases.

Some effects of higher aromaticity in EPDM compounds have been found to be related to the type of curative system [5]. Intermediate and high viscosity EPDM rubbers in a black-loaded, peroxide-cured stock were tested with oils of increasing aromatic content. Although tensile strength remained about the same, elongation and compression set increased significantly. It can be assumed that the increasing number of double bonds in the oils are competing with the polymer for the curative. When EPDM was tested with a sulfur donor system, there seemed to be little correlation of results with aromatic content. In the same paper aromatic and paraffinic oils were tested at 50 and 200 parts loading. At the higher loading, compared to the paraffinic oil, the aromatic oil resulted in lower tensile strength, modulus, and hardness. Compression set results showed no correlation with aromaticity.

Highly naphthenic oils have excellent color and heat stability. They are frequently used with EPDM compounds. Paraffinic oils show excellent resistance to discoloration by UV light and have good low temperature flexibility. However, they simply do not have the compatibility to serve well in nitrile, neoprene, or SBR stocks.

Given the number of factors involved, any assessment of which type of oil is best for each polymer is necessarily subjective. However, such an assessment is made in Table 13.2.

If a rubber compound is being used in direct or indirect contact with food, the mineral oils in the compound must meet the quality requirements of FDA regulation 178.3620(c) for such oils. This regulation applies, for example, to rubber articles for such repeated uses as sorting and inspection belts in canneries. Adhesive coatings and compounds used for packages containing food constitute another application. Since these regulations change from time to time, usually by expanding, compounders should rely on their suppliers for up-to-date information.

Table 13.2 Petroleum Oil Section Chart for Polymers

Polymer	Oil suitability		
	Paraffinic	Naphthenic	Aromatic
Natural rubber	Excellent	Fair	Fair
SBR	Fair	Good	Excellent
Butyl	Excellent	Good	Poor
Polybutadiene	Fair	Good	Excellent
Neoprene	Poor	Good	Excellent
Nitrile	Poor	Fair	Excellent
EPDM	Excellent	Good	Good

III. PINE PRODUCTS

There has been a long-term decline in production of materials extracted from pine trees, at least in the United States, one of the world's leading producers. The products have stiff competition from other materials—often petroleum-based. The industry is labor-intensive, since the pine trees are tapped for the gum, or pine stumps are collected and steam-distilled. As with tapping for rubber latex, such work is rather unattractive, and mechanization of the process is difficult. There is a possibility for increased production, at least for terpene resins. Citrus fruit skins contain significant amounts of pinenes (the monomer used), and if an economical commercial process could be worked out for their retrieval at citrus-processing plants, the supply picture would improve.

Pine products start with the fractional distillation of pine tree tappings to give turpentine. This is gum turpentine; if pine tree wood stumps are steam-distilled, the product is wood turpentine. Both turpentines contain appreciable quantities of terpenes, which are oligmers of isoprene and have the general $(C_5H_8)_n$.

Terpene solvents can be made from turpentine. In rubber the high solvency terpene hydrocarbon fractions like dipentene are used as reclaiming agents for natural and synthetic rubbers. With flash points as low as 115–118°F (Tag Closed Cup), mixing, especially in internal mixers, could be hazardous. Terpene solvents are near colorless, have a pleasant terpene odor, and in many cases comply with FDA requirements for use in food packaging. At the high end of the distillation range is a product like Hercules Inc.'s Hercoflex 600. This pale, high boiling liquid is recommended for nitrile and neoprene stocks. An efficient plasticizer at low temperatures, it has low plasticizer loss after heat aging and retains properties well.

Pine tar, a residue from the distillation of pine gum, is an efficient softener, especially for natural rubber. It is rarely used at levels higher than 5 parts.

Besides its softening action it helps in the dispersion of carbon black, has age-resistant qualities, and enhances the property known as building tack. With good building tack, hand-made built-up items such as expansion joints hold their shape well before curing. Because of its dark color, pine tar cannot be used for white or light-colored goods. It does have a slight retarding effect on cure.

Terpenes, usually β-pinene, can be polymerized to give resins that have special niches in compounding. These resins can vary from viscous liquids to hard, brittle materials. Some are water-white. They are thermoplastic and very tacky when soft. Molecular weight is low, from approximately 550 to 2200, and is a mixture of various molecular weights. They are stable to heat and UV radiation.

Terpene resin features that are of use in compounding include wide compatibility with NR, SBR, IIR, CR, petroleum hydrocarbons, vegetable oils, and waxes. The more liquid resins are excellent tackifiers. Because they are nontoxic, nonphytotoxic, and nonsensitizing to the skin, they can be used in compounds that come in contact with food products. Terpene resins are used as a base for chewing gum.

IV. SYNTHETIC RESINS

The remaining synthetic resins used in rubber are largely coumarone indene (CI) resins and petroleum hydrocarbon resins, with the former being supplanted by the latter. As in the case of the terpene resins, coumarone indene resin production is limited by the scarcity of feedstock. These resins are made from heavy-solvent naphtha obtained from the distillation of coal tar. The latter in turn is a by-product of coke production and, with diminished production of steel in fewer coke ovens, at least in the United States, fewer such units are in operation.

Heavy-solvent naphtha is rich in coumarone and indene, especially indene. These unsaturates were formerly polymerized by sulfuric acid, but modern practice is to use BF_3 or BF_3 etherates, which produce lighter resins. The catalyst can be removed by an alkaline wash or lime after polymerization. The resin can be isolated by steam distilling off the unreacted naphtha. Coumarone indene resins are largely polymers of indene and have relatively low molecular weights, are light in color, and have good color stability. Their specific gravity ranges from 1.01 to 1.14, and they have softening points from 10 to 155°C by the ring-and-ball method (ASTM E28).

Unlike rubber processing oils, neither coumarone indene resins nor other resins are used primarily to soften the mix, rather, they act as plasticizers, help in the dispersion of pigments, and hold stress–strain properties at high levels. It is important that mixing temperatures be high enough to melt the resin if the higher grades are used. With the softening point range given, this problem should not arise.

High strength white or light-colored SBR compounds are difficult to formulate. Here coumarone indene resins are especially valuable, allowing satisfactory stocks to be made with mineral fillers. This property can be extended down to lower quality stocks. For example, a suggested washing machine hose formulation in the literature [7] uses 20 parts of a CI resin with 215 parts of white fillers on 100 parts of SBR. The writer has found CI resins particularly useful in providing building tack in otherwise dry stocks by using small dosages, say 3–10 parts of the lower melting points resins.

Because production of CI resins is limited by the availability of feedstock, petroleum hydrocarbon resins are replacing them in most areas. Raw material for these low molecular weight thermoplastic resins are cracked petroleum fractions. The unsaturated monomers from the fractions are again polymerized using BF_3 or BF_3 etherate as a catalyst. When the reaction is complete, the catalyst can be deactivated with water, aqueous alkalies, or lime before being removed. Some hydrocarbon resins can be made lighter by treatment with decolorizing clay.

Hydrocarbon resins are pale yellow to dark brown and range from liquids with softening points below 10°C to hard, brittle resins with softening points from 180 to 190°C. They come in two main types depending on the feed source. The C_5 fraction gives an aliphatic olefin resin, rather dark in the Gardner scale at 11, with a specific gravity of 0.98. The other type is aromatic resins which uses a C_9 feedstock with no C_5 material present. Specific gravity ranges from 1.05 to 1.08. This latter type of feedstock has monomers that have been found in heavy-solvent naphtha, so that some of the same processing technics can be used.

Hydrocarbon resins are used in a wide variety of rubber products, the dosage rarely exceeding 20 phr. The aliphatic types are mainly used as tackifiers. Functionally modified aromatic resins or aromatic modified aliphatic resins can be used not only as tackifiers but as reinforcing resins. Particular attention should be paid to resin color and the color stability of the resin if light-colored or white goods are being produced.

One group of synthetic resins should be mentioned, even if briefly, because of their special properties. These are the phenolic resins. Using an acid or base as a catalyst, they are made from phenol, formaldehyde, cresols, and higher aldehydes. The range of possible products is increased by including olefins, which can react with phenol using a Friedel–Crafts catalyst.

The chemistry of these resins can become very complex and is not reviewed here. In general these resins are compatible with NR, SBR, CR, NBR, and BR rubbers. In dry rubber compounding they serve as reinforcing resins, as thermoplastic tackifiers and processing aids, and as resins for use in adhering rubber compounds to fabrics.

A leading producer is phenolic resins, the Occidental Chemical Corporation, markets its line as Durez resins. Although the following examples specify a Durez resin, one could expect much the same effects with equivalent grades of phenolic resins made by other producers.

Phenolic resins are very compatible with nitrile rubbers and are used to reinforce nitrile compounds. In a simple nitrile compound, 50 parts of Durez 12687 resins showed a tensile strength of 3100 psi, an elongation of 300%, and a hardness of 94 against 2050 psi, 350% elongation, and only 65 Shore A hardness for a 60-part SRF carbon black compound [8]. Phenolic resins serve well in tackifying dry stocks like butyl. With only 5 parts of Durez resin 31671 added to the mix, the before-and-after tack improvement was significant. The quick grab, tack-and-dwell tack values (Instron tack test) went from, respectively, 1.0 and 1.6 lb/in. to 3.3 and 5.9 lb/in.—over three times as much.

V. ESTERS

Esters are formed by the condensation of an acid with an alcohol and the splitting out of water. They are identified by giving first the group derived from the alcohol and then the group from the acid (e.g., dioctyl adipate). Esters are relatively high priced and are rarely used except for a specific purpose—to give low temperature flexibility to more polar rubbers like neoprene, butadiene, and acrylonitrile. They are infrequently used in amounts over 30 pph. Common ester plasticizers are dioctyl adipate, dioctyl sebacate, and butyl oleate.

There is a very wide variety of ester plasticizers with widely different properties. Some characteristics to be checked before adding a particular plasticizer to a compound in service include compatibility with the polymer, effectiveness in improving the low temperature properties, volatility, effect on cure rate, and resistance to ozone. Some plasticizers impair fungus resistance in a compound.

Many of the esters used in rubber compounding will be extracted if in contact with oils like ASTM no. 1 or no. 3. If the amount of extraction is excessive, the addition of polyesters can help reduce this extraction.

VI. NATURAL FATS AND OILS

Vegetable oils and fats, apart from pine products, are occasionally used in rubber compounding. Certain unsaturated oils such as linseed, rapeseed, and safflower oil are used in neoprene when special effects are needed. These effects include low temperature flexibility, tear resistance, and antiozonant protection. As yet there appears to be no good nonstaining antiozonant; linseed oil can protect light-colored neoprene stocks from this kind of attack without staining. There is a difficulty with nonpetroleum softeners not mentioned earlier, however. Rubber compounds that serve in moist, warm areas like the tropics can be impaired or even made useless by fungus attack. A prime source of nutrient for the fungi appears to be in the softener or plasticizer of rubber and plastic compounds. As such, an unsaturated oil like linseed should not be used in such service conditions.

Allied to vegetable oils are the vulcanized oils such as factice, commonly known as white or brown "subs," or substitutes. At the time they were invented, solid vulcanized oils, factice, were considered substitutes for rubber itself. A large variety of unsaturated animal and vegetable oils can be treated with sulfur or sulfur monochloride (S_2Cl_2) to give a rubbery product. A familiar example of this is the art gum eraser sold in stationery stores. If the oil is vulcanized with sulfur, the product is dark and is called brown subs. If sulfur monochloride is used, the product is light-colored and is referred to as white subs.

Three properties of vulcanized oils give them value in rubber compounding: they are not thermoplastic, they flow well under pressure, and they can absorb large amounts of liquid rubber plasticizers. The nonthermoplastic nature of vulcanized oils helps uncured forms like tubing retain their shape and dimensions better. Their fluidity under pressure does help in dispersing other ingredients in the mix, serving as a dry plasticizer, as it were. Their ability to absorb large amounts of plasticizers allows the incorporation of larger amounts of plasticizer than would be possible otherwise and yet the plasticizer doesn't exude after curing.

Because of these characteristics, factice is largely used with synthetic rubbers to make soft rubber products—mainly items whose hardness is under 40 Shore A. For example, a neoprene stock with a Shore A durometer reading of 10 can be made by using 100 parts of oil and 50 parts of factice.

Two other uses of factice might be mentioned. Frequently these solid, vulcanized oils are used in pressure-sensitive adhesives, where they prevent hot and/or cold flow during storage. Erasers are usually made of 2–3 parts white factice and 1 part rubber. A rubber compound itself would remove graphite pencil marks, but the discoloration would remain on the rubber, preventing further use. With factice, the compound abrades and a fresh surface is presented.

REFERENCES

1. Ludwig et al. *India Rubber World, 53*, 173 (1945); *112*, 731.
2. Files, E., *A Basic Look at Fluid Effects on Elastomers*, Rubber and Plastic News, June 10, 1991.
3. Stephens, H. L., Plasticizer Theory—An Overview Based on the Works of Doolittle, Bueche, and Ferry, Educational Symposium No. 10. ACS Rubber Division, May 1983.
4. Dimeler, G. R., Plasticizers for Rubber and Related Polymers, Educational Symposium No. 9, ACS Rubber Division, October 1982, p. 36.

5. Godial, M. J. EPDM Plasticiz n.g, Educational Symposium No. 10, ACS Rubber Division, May 1983.

6. Dimeler, G. R., Plasticizers for Rubber and Related Polymers, Educational Symposium No. 9, ACS Rubber Div sion, October 1982, p. 76.

7. Williams, E. E., in *The Vanderbilt Rubber Handbook*, 12th ed., Vanderbilt, Norwalk, CT, 1978, p. 721.

8. Durez Phenolic Resins for the Rubber Industry, bulletin from Occidental Chemical Corp., Durez Resins and Molding Materials, 673 Walck Road, North Tonawanda, NY 14120-3493.

Age Resisters

I. INTRODUCTION

The term *age resister* is an umbrella term covering materials whose addition prolong the useful life of a rubber or rubber product. The most common examples are antioxidants and antiozonants, designed to inhibit oxidative and ozone-caused deterioration, but ultraviolet light protectors and antiflex agents are included as well. Age resisters added to raw synthetic polymers are usually called stabilizers. Some ingredients, such as carbon black, give significant age resistance but are not so classified because they are not added for that purpose. Also excluded from this group are coatings, materials like liquid polyurethane applied to such products as automotive mounts to protect them from oil drips, ozone, etc. This chapter deals with antioxidants and antiozonants.

The results of oxidative attack depend on the polymer. Some rubbers, such as natural rubber, become soft and sticky; others, such as SBR, become hard. Ozone attack is manifested by cracking at the surface perpendicular to the stress. Ozone is destructive to stretched rubber, not unstretched rubber, and has little effect on rubbers with a saturated backbone like EPDM. For each compound there is a critical elongation, which must be reached before ozone attack is significant. The minimum elongation at which cracking can occur appears to vary with the polymer. Natural rubber can crack at elongations as low as 3%, but butadiene acrylonitrile cracks at 8%, neoprene at 18%, and butyl at 26% [1]. Ozone cracking is proportional to ozone concentration, and rather low levels of concentration are needed to cause deterioration—usually not over 1 pphm (10×10^{-8}) in air [2].

The most dramatic evidence of perishing by ozone was the deterioration of tires in the Los Angeles area some years ago when smog and the associated ozone were at very high levels. On some occasions ozone levels reached 100 pphm of the air there and at other locations (but less frequently) in the United States and Canada. With the environmental controls now in place and being expanded, such levels are rare, but the continuing presence of ozone will require rubber compounds to have antiozonant protection in most applications.

Age resisters were not used in rubber products prior to 1920, although such natural materials as pine tar pitch with age resistance properties were used in mixes for other purposes. C. C. Davis, one of the most prominent rubber chemists of his time, wrote in 1932 that good aging stocks could be had using low sulfur and sulfurless cure systems [3]. Furthermore, he believed that age resisters were unneeded if such stocks were properly cured. Few compounders followed his advice. By 1955 it was estimated that usage of age resisters was well above 1% of the polymers used. This relative usage has probably increased since then.

For example, windshield wiper compounds, which must withstand oxygen and ozone attack abetted by sunlight, high summer temperatures, and frequent flexing, contain 3–5 parts of age resister on 100 of polymer. The judicious use of age resisters can probably prolong the service life of an article 2–4 times.

Although much research has been directed to studying the mechanisms by which oxygen and ozone attack rubber, there are still areas of uncertainty. In the early 1960s, it was found that traces of peroxides can be detected even in very carefully prepared diene rubbers. Their presence explained why elastomers could start deteriorating at relatively low temperatures. Other pro-oxidants for rubbers are UV light and certain copper and manganese salts (although many such salts do not affect rubber). The work continues and 1981–2 research by Keller and Stephens [4, 5] shows that unsaturated fatty acids can accelerate NR oxidation. A brief review of theories on oxygen and ozone attack might be useful for orientation.

II. THEORIES ON OXIDATIVE AND OZONE ATTACK

It is generally believed that the oxidation of elastomers follows the kinetic mechanism conceived by Bolland and associates [6] to illustrate the oxidation of low molecular weight hydrocarbons. The scheme, outlined in Figure 14.1, consists of three steps: initiation, propagation, and termination. The deterioration process starts with the formation of free radical species. The polymer represented by RH can break down during manufacture from stresses in baling or high temperatures in drying to form the initiation process. As mentioned earlier, this

(a)   RH $\longrightarrow$ R• + H•

(b)   R• + O$_2$ $\longrightarrow$ ROO•

ROO• + RH $\longrightarrow$ ROOH + R•

2ROOH $\longrightarrow$ RO• + ROO• + H$_2$O

ROOH $\longrightarrow$ RO• + •OH

(c)   2R• $\longrightarrow$ R—R

R• + ROO $\longrightarrow$ ROOR

2ROO• $\longrightarrow$ (x)

Figure 14.1 Oxidation of elastomers: (a) initiation, (b) propagation, and (c) termination (x-stable products).

breakdown can occur at relatively low temperatures because of the presence of small amounts of peroxides.

Once the polymer chain has been cleaved, the rubber radical can react with oxygen to form a rubber peroxide. This in turn can attack another chain to form a rubber hydroperoxide and a rubber radical. Other reactions as shown help in this autoxidation process. If at this stage the material is exposed to UV light, free radicals are produced by the breaking down of hydroperoxides and carbonyl compounds. Termination of these reactions can occur when two rubber radicals unite to form crosslinked polymer. A rubber radical can unite with a rubber peroxide to form another type of crosslink. Finally, two rubber peroxide radicals can unite to form stable products.

Ozone attack on rubber is somewhat different. Ozone attacks only rubber that has unsaturation in the backbone chain. Although it does attack unstretched rubber, cracks are formed only when stretched rubber is exposed, and this is the deterioration of prime concern. Ozone attack on unstretched rubber ceases when all the surface double bonds have been used up. The reaction is rapid at first, chokes off, and ultimately leaves a gray film on the rubber. This process is called frosting. Ozone can attack saturated hydrocarbons and sulfur crosslinks, but the rates are low. Because of this, ozone can attack polysulfide rubbers, although the latter have no double bonds. Further information is given by Kendall and Mann [7].

Figure 14.2 shows the steps believed to occur in this process. With ozone, the target is the double bonds in diene rubbers. The resultant product—a bridging of one side of the double bond with both ends of the ozone molecule to form a cyclic ozonide—is called a molozonide. This unstable compound breaks down

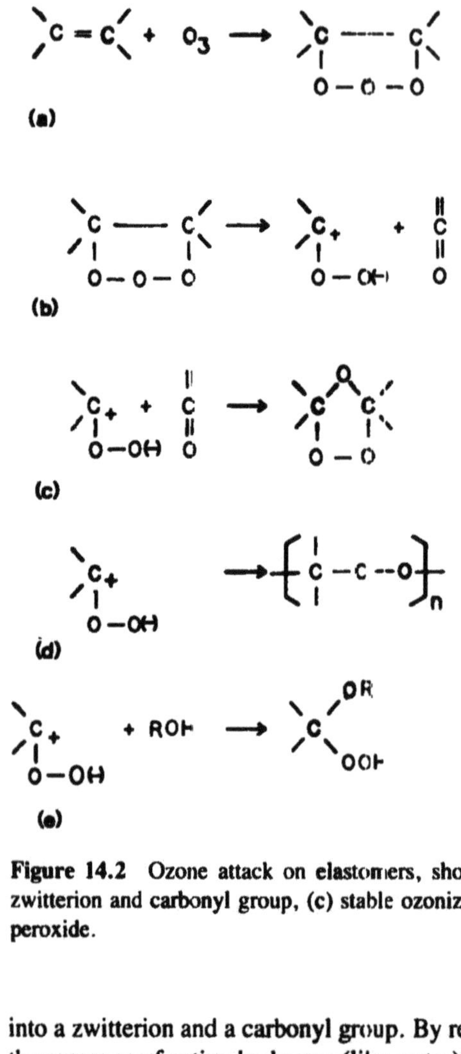

$$>C = C< + O_3 \longrightarrow >C \cdots C<$$
$$O - O - O$$

(a)

$$>C \longrightarrow C< \longrightarrow >C_+ + \overset{\parallel}{C}$$
$$O - O - O \qquad O - OH \quad O$$

(b)

$$>C_+ + \overset{\parallel}{C} \longrightarrow >C \overset{O}{\underset{O-O}{\diagdown}} C<$$
$$O - OH \quad O$$

(c)

$$>C_+ \qquad \longrightarrow \left\{ \overset{\mid}{C} - C - O \right\}_n$$
$$O - OH$$

(d)

$$>C_+ + ROH \longrightarrow >C\overset{OR}{\underset{OOH}{\diagdown}}$$
$$O - OH$$

(e)

Figure 14.2 Ozone attack on elastomers, showing formation of (a) molozonide, (b) zwitterion and carbonyl group, (c) stable ozonize, (d) polymer peroxide, and (e) hydroperoxide.

into a zwitterion and a carbonyl group. By recombination of the latter two and in the presence of active hydrogen (like water), one of three products is possible: a relatively stable ozonide, a polymeric peroxide formed from the carbonyl oxide, or a hydroperoxide. These conclusions have been supported by experiments that showed, for example, consumption of double bonds and the formation of ozonide. With stretched rubber, the zwitterion and the aldehyde fragments separate, the ozonides and peroxides form remote from the starting crack, and fresh rubber chains are in contact with ozone. As the rubber chains keep on breaking, they retract in the direction of the stress and expose underlying unsaturation. As this process continues, the characteristic ozone cracks are formed.

III. EFFECTS OF OXYGEN AND OZONE ATTACK

The kind of deterioration caused by oxygen depends on the nature of the polymer involved. In natural rubber and butyl, oxygen causes chain scission. This shortens the polymer chains and in the early stages leads to softness in the rubber; as deterioration proceeds, it becomes sticky. This can often be noted in stationers' rubber bands, usually made of natural rubber, which have been exposed to sunlight. In contrast, SBR, CR, BR, and EPDM harden and become leatherlike as the oxygen crosslinks the polymer chains to form stiffer compounds.

Ozone attack is much more rapid than oxygen attack and is essentially a surface phenomenon. The result is visible, tiny cracks perpendicular to the stress. If the elongation is well above the critical range, cracks are numerous but small. As the elongation decreases, the cracks beome fewer but deeper. As the cracks grow deeper, rubber product failures increase. Some compounds can have both antioxidant and antiozonant properties.

IV. ANTIOXIDANT AND ANTIOZONANT TYPES

A. Antioxidants

Before reviewing how the effects of oxygen and ozone on rubbers are impeded by using age resisters, it is useful to note the materials available. Age resisters have yet to receive the extensive classification that carbon blacks, for example, have been given. Most current popular antioxidants can be separated into amines, phenols, and thioesters. Four amine classes have been, or are, widely used. These are:

1. Naphthylamines
2. Diphenylamine derivatives
3. Paraphenylenediamines
4. Dihydroquinolines

Until fairly recently both α- and β-phenyl naphthylamine were widely used because both were effective antioxidants. At least in the United States, their use has decreased as manufacturers stopped marketing these products on the grounds of possible toxicity. Structures for the four types are shown in Figure 14.3. Typical commercial brands in these classes are Nonox A (phenyl-β-naphthylamine) from Imperial Chemical Industries Ltd. (ICI), Wingstay 29 (styrenated diphenylamine) from Goodyear Chemicals, Santoflex IP (*N*-isopropyl-*N'*-phenyl-*p*-phenylenediamine) from Monsanto, and Agerite Resin D (polymerized, 2,2,4-trimethyl-1,2-dihydroquinoline) from R. T. Vanderbilt.

The main phenolic antioxidants consist of five types. These are:

1. Hindered phenols
2. Hindered bisphenols

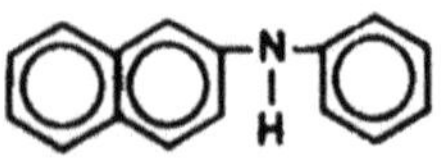

(a)

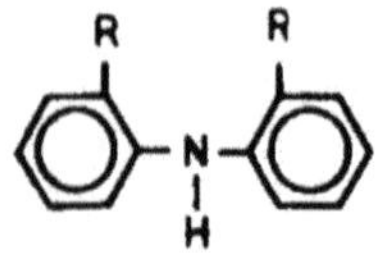

(b)

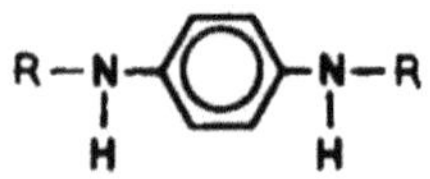

(c)

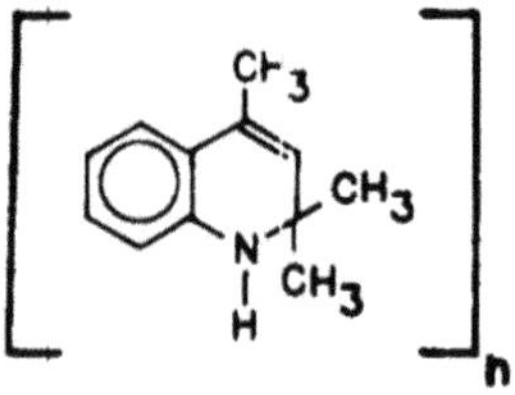

(d)

Figure 14.3 Amine antioxidant types (a) phenyl-β-naphthylamine, (b) diphenylamine, (c) p-phenyldiamine, and (d) dihydroquinoline.

3. Hindered thiobisphenols
4. Polyphenols
5. Polyhydroxyphenols

Structures of common antioxidants in these classes are shown in Figure 14.4.

Although not used in rubber compounding, perhaps the most popular antioxidant worldwide is 2,6-di-*tert*-butyl-*p*-cresol, frequently called BHT. This widely used antioxidant in food products is mentioned here because its structure is similar to that of phenol antioxidants used in rubber.

Hindered phenols are monophenols, variants of the BHT model. The bisphenols are two hindered phenols hooked together. If the linkage is not a hydrocarbon group but sulfur, the product is called a hindered thiobisphenol.

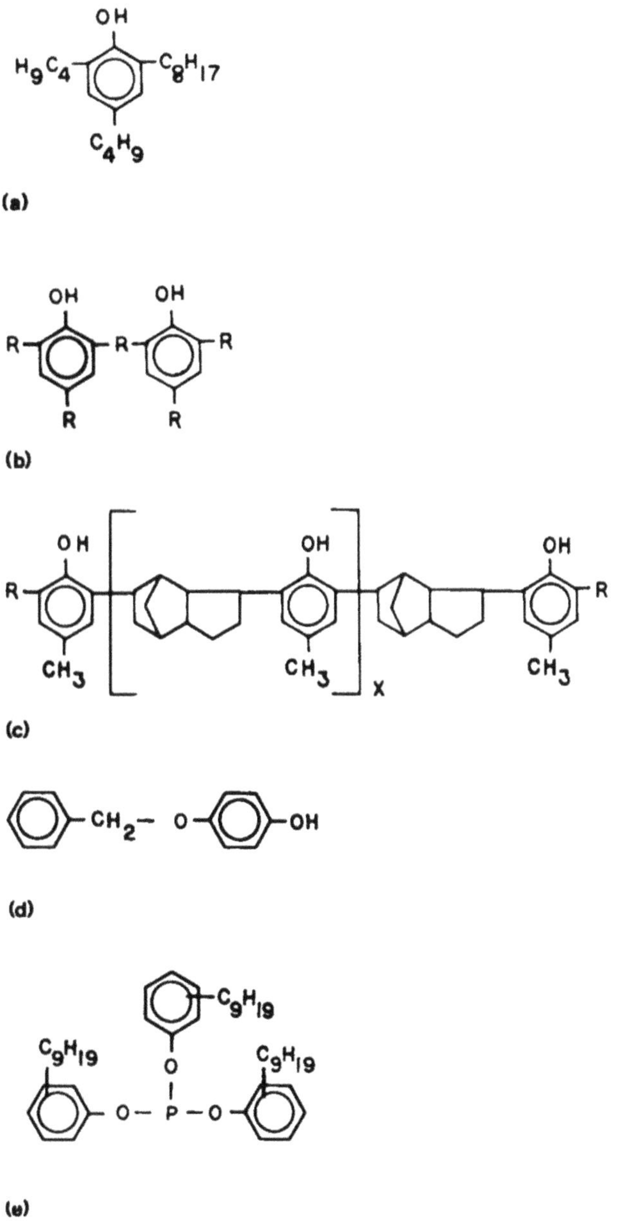

(a)

(b)

(c)

(d)

(e)

Figure 14.4 Common phenolic antioxidants: (a) hindered phenol, (b) hindered bis-phenol (hindered thiobisphenols have sulfur instead of R bridge), (c) polyphenol (Good-year's Wingstay L), (d) polyhydroxyphenol, and (e) phosphite antioxidant Polygard (Goodyear Corp).

Polyphenols are similar to bisphenols, but the molecule has been enlarged and has a much greater molecular weight. Lower volatility, a desirable property, is usually associated with higher molecular weight. Polyhydroxyphenols, derivatives of hydroquinone, are used for protecting uncured rubber stocks such as adhesives and, more important, as a stabilizer in synthetic rubber production.

Phosphite antioxidants are infrequently used in compounding; however, they do serve as stabilizers in synthetic rubber manufacture. A widely used antioxidant of this kind is tris(nonylphenyl)phosphite, marketed by Goodyear Chemicals under the brand name Polygard (Fig. 14.4d).

Protection against ozone can be obtained by using wax or an ozone-resistant rubber. The latter was referred to earlier; both approaches are discussed later in this chapter.

V. ANTIOXIDANT AND ANTIOZONANT PROTECTIVE MECHANISMS

It would appear that practice in providing good oxygen and ozone protection is in advance of our understanding of how that protection works. Total antioxidant action ordinarily entails three functions. First is the absorption of UV light, which catalyzes oxidation; this is usually done anyway by the fillers in the rubber. If the stock does not have pigment and is subject to outdoor exposure, specific UV absorbers can be added. Second is the decomposition of initiating peroxides, which are converted into nonradical products. It is believed that mercaptans, thiophenols, and other organic sulfur compounds function in this way [7]. Finally, we have the stopping of the radical chain reactions.

Antioxidants of the phosphite type used as stabilizer can neutralize the hydroperoxides:

$$ROOH + HA \rightarrow stable\ products$$

where HA represents the antioxidant.

The phenols and amines used by the compounder stop the chain reaction by interposing themselves in it.

$$ROO \cdot + HA \rightarrow ROOH + A \cdot$$

The resultant free radical can link up with a like radical to form a stable product or react with another peroxide radical:

$$ROO \cdot + A \cdot \rightarrow ROOA$$

The way chemical antiozonants protect rubber is not entirely clear. Current thought invokes a scavenging action in combination with the formation of a protective film [8]. According to the scavenging theory, the antiozonant migrates to the rubber surface, where it scavenges the ozone present because of its reactivity toward that gas. Several papers have reported on this process [9–12]. Unlike antioxidants, antiozonants react more readily with ozone than with the double bonds in rubber, so the latter is protected.

The second mechanism believed to be involved is the formation of a protective film on the rubber surface. Such a film has been analyzed and found to consist of unreacted antiozonant and its ozonized products. Furthermore, if the film is removed by a solvent, the rubber loses its resistance to ozone attack. Support for the protective film theory is evidenced in the work done by Andries and associates, who studied the surface of unloaded and carbon black loaded natural rubber stocks [13]. These investigators found that the antiozonant bloomed to the surface, where it both reacted preferentially with the ozone and formed a flexible film of ozonide products that protected the substrate.

VI. USING AGE RESISTERS IN COMPOUNDING

Before dealing with the use of age resisters already described, we present two other methods of providing protection.

Paraffin wax is a cheap and effective antiozonant, usually used in conjunction with a chemical antiozonant, the latter giving protection in conditions of dynamic stress. There is probably much ignorance about the composition, mode of action, and proper selection of this widely used age resister, however.

By oversimplifying the way paraffin works can be described quickly. When added to the compound it usually dissolves at mixing temperatures; if not then, during vulcanization. Paraffin wax has limited solubility in rubbers, so on cooling precipitates out and migrates to the surface. There it forms a thin chemically inert film, putting up a physical barrier to the ozone, which would attack the rubber. If through movement of the rubber article, scuffing, or washing, the film is scraped or flaked off, the rubber is subject to attack until another layer forms.

Ferris, et al. [14] pointed out that the performance of an antichecking or ozone-resistant wax controlled by its chemical composition. Paraffin waxes produced by solvent extraction from the lubricating oil fraction of petroleum consist almost entirely of normal paraffins (alkanes) with a small proportion of slightly branched chain paraffins. Both groups have the formula $C_nH_{2n} + 2$, with n stretching from 18 to 50. The waxes so produced have melting points ranging from 52 to 68°C and are usually sold as grades like 52°/54°, which represents a mixture of components having a melting point between 52 and 54°C.

There is considerable overlap in composition between one grade and another. With straight-chain molecules like the paraffins the waxes are macrocrystalline. If higher boiling lubricating distillates are solvent-extracted, the so-called amorphous waxes are obtained. These are much higher in molecular weight and have many more branched chains. With this bulk they do not crystallize easily and are called microcrystalline or amorphous waxes. There is no precise dividing line between the lowest melting point paraffin wax and the highest melting point wax to separate the macrocrystalline material from the microcrystalline. Wax blenders may use up to 15% of the higher melting point waxes, the amorphous waxes, to give long-term protection. With its high molecular weight and very branched structure, paraffin wax has a very slow rate of migration.

Thus in reality a paraffin wax for rubber cannot behave like a single compound; rather, it is like a large number of different hydrocarbons, all having their own diffusion characteristics. As the rubber vulcanizate cools, the wax is in a supercooled condition and crystallizes out, migrating to the surface because of the concentration gradient between the surface and the interior. At the surface the film forms as an amorphous or microcrystalline mass because of the small amount of branched material preventing the packing of the paraffin mass in a crystalline form. Diffusion takes place until the wax concentration in the rubber is at the solubility limit for that temperature. The diffusion rate is affected by the solubility of the wax in that particular rubber, the fillers—their kind and loading—and especially the temperature. According to Jowett [15], the temperature variations found during outdoor exposure may affect wax diffusion coefficients many thousand times.

For a given wax in a rubber compound there is a temperature at which migration is at a maximum. This of course is an average of all the individual fractions in the mix. At low temperature such as 0°C only the low carbon number fractions such as C_{18}–C_{21} have diffusion rates that provide surface protection at a reasonable level. At high temperatures (50°C), the lower carbon fractions are soluble in the rubber and do not migrate; only the high carbon fractions like C_{33} and up will diffuse.

From the foregoing it is seen that the job of selecting a wax for a product is not a simple one. For a product that will be used at or around room temperature, most waxes will do the job. However, if temperature extremes (e.g., 0 or 50°C) will be experienced for any length of time, during which the film may be flaked off, scraped, buffed, cleaned, or otherwise removed, more care is needed. The wax will have to be blended in proportions that will ensure the availability, at the temperature expected, of a sufficient supply of C number fractions that diffuse readily at such temperature. For example, if 0°C is anticipated, there must be enough C_{18}–C_{21} fractions to provide protection.

To make right choices, then, the compounder should learn from the supplier the carbon number distribution and the ratio of straight-chain to branch-chained hydrocarbons. With this information and by testing at use temperature with a test piece designed to show where the critical elongation of the compound is, the compounder can be confident that a given wax is suitable.

One way to protect rubber from ozone attack is to add an ozone-resistant rubber to it. The method has limitations, but there is no other way to protect the rubber under dynamic conditions, without staining. Usually the ozone-resistant polymer is used at 20–50% of the total polymer weight. The protection effect is due to the existence of the polymers in separate dispersed phases in the compound. A crack that starts in the weaker polymer may travel to the domain of the other polymer. Here the stress is reduced at the crack tip to less than the critical energy required, and the crack ceases. If, there is for example, only 10% of ozone-resistant polymer in the compound, the globules are small in the mix and the crack travels around them. No improvement is achieved above the 50% loading.

The most common usage of this technique is in adding 20–40 parts of EPM or EPDM to NR, SBR, or BR. The ozone resistance of NBR can be improved by adding EPDM or CR. The disadvantage of this method is the frequently poorer properties of the blend.

There is a rather bewildering array of chemical age resisters from which the compounder can choose. It is important to know the atmosphere in which the product will perform as specifically as possible. Generally the choice of antioxidant is made first and then, if it is needed, an antiozonant is selected. Some antioxidants (e.g., the *p*-phenylenediamines) can serve as antiozonants as well. The dosage of age resisters is usually 0.5–5 phr; most formulas call for 1–3 phr. Severe or unusual service conditions can make a mockery of such averages. One massive natural rubber automotive part used 8–12 parts of a resinous hydroquinoline antioxidant successfully. Not only was high heat resistance required, but it was believed the antioxidant served as a plasticizer and facilitated post-vulcanization metal bonding.

The choice of antiozonants compared to antioxidants is small. Few chemicals react rapidly with the normal concentrations of ozone. To protect NR, SBR, BR, and NBR rubbers, one must rely on N,N^1-disubstituted *p*-phenylenediamines (PPDAs). The three basic types are N,N^1-dialkyl-, N-alkyl-N^1-phenyl, and N,N^1-diaryl-. One strong antiozonant provided by several suppliers is N-1,3-dimethylbutyl)-N^1-phenyl-*p*-phenylenediamine:

With a crystallizing point range of 44–50°C, this PPDA can be obtained as a liquid in heated tank cars, as flakes in kegs, or as a solid in metal drums. This antiozonant gives good protection under dynamic as well as static ozone attack; it also offers protection against flex cracking and heat aging. For these reasons it is frequently used in belts and tires. In the case of tires there is a further advantage, namely its resistance to water leaching. Again there are tradeoffs—this additive discolor compounds and causes in gration stains.

By and large. properties considered by a compounder are much the same whether an antioxidant or an antiozonant is being chosen. These properties include the following, not necessarily in order of importance:

1. Physical form
2. Amount of protection
3. Persistence
4. Solubility
5. Toxicity
6. Color, tinting, and staining propensities
7. Cost

Age resisters (also called antidegradants) are manufactured as liquids, viscous liquids, semisolids, powder, flakes and rods. For ease in weighing and general handling, liquids or powders, flakes, and rods are preferred. Semisolids cannot be handled well on automatic weighing equipment and are somewhat troublesome in batching up by hand. The amount of protection varies greatly with the particular materials used. For example, after 2 days of oven aging at 100°C a phenolic phosphite protected natural rubber compound lost three times as much tensile strength and twice as much elongation as one protected by a phenyldiamine type.

Antioxidants and antiozonants tend to disappear from a compound, so obviously more protection is afforded if they are highly persistent. A major loss is due simply to volatilization, which depends on the temperature of service, circulation of air around the product, etc. Higher molecular weight products generally have lower volatility. but type plays a significant role, hindered phenols being much more volatile than amines. When antioxidant is removed by direct oxidation rather than as a chain stopper, its effectiveness is decreased.

The solubility of these materials is of concern in two ways. If the dosage in the compound exceeds its solubility in the polymer it will bloom. This unsightly defect can impair the use of the product. Some antioxidants do not bloom but others can bloom at concentrations as low as 0.35 phr. Besides solubility in the rubber, solubility in the environment to which the rubber product is subjected is important. For example, when rubber used in frequently washed items, such as elastic foundation garments, the antioxidant may leach out, and the article wear out sooner. Hopefully, the relatively recent development of polymer-bound antioxidants that do not have this fault will be considerably expanded.

Toxicity is a matter of importance in using any compounding material but is of more concern with age resisters and accelerators than with materials like sulfur and stearic acid. Material safety data sheets, available from most manufacturers, suggest correct handling procedures. If the rubber product comes in contact with food, then regulations of the U.S. Food and Drug Administration apply.

An antioxidant can affect the color of the rubber into which it is mixed or even an adjacent rubber part. If an antioxidant tints a light-colored stock into which it is mixed, the effect may not appear until the item has been exposed to light for some time. Obviously such effects do not matter if the stock is black. This tendency to discolor can result in adjacent rubber surfaces, or other surfaces like white refrigerator shells, being stained (the so-called contact stain). At times this staining tendency is strong enough to migrate through an adjacent rubber compound. The *p*-phenylenediamines seem to be especially prone to discoloring and staining effects. In most cases phenolic antioxidants are better than amines.

Cost is always a consideration in rubber compounding and especially with age resisters (which in most cases constitute the most expensive ingredient on a pound basis going into the compound). Cost can sometimes be reduced by using representatives from two groups, since synergistic effects often result and the total poundage is reduced. Inasmuch as the law of diminishing return applies, (i.e., increasing the amount of age resister does not proportionately increase protection), a lower dollar cost may be obtained with a smaller amount of a very effective agent than a greater amount of a less effective one.

The age-resistant properties of compounds are largely tested by accelerated test such as hot air aging, measuring the rate of crack growth in strips subjected to constant flexing, and visual appearance of stretched samples in high ozone chambers. Although helpful, these tests cannot be expected to correlate closely with actual results from service life. Constant outdoor aging is recommended, and if a faster indication is wanted, consideration should be given to using the facilities of environmental testing services in such areas as Florida and Arizona. Both locations give high ambient temperatures and maximum sunlight exposure, but Florida has high humidity, and Arizona a desert-dry atmosphere.

REFERENCES

1. Edwards, D. C., and Storey, E. B., *Rubber Chem. Technol.*, 28, 1096 (1955).
2. Friberg, C., *Rubber Age*, 106, 341 (1974).
3. Davis, C. C., *Vanderbilt News*, 2(3), 48 (1932).
4. Keller, R. W., and Stephens, H. L., *Rubber Chem. Technol.*, 54, 115 (1981).
5. Keller, R. W., and Stephens, H. L., *Rubber Chem. Technol.*, 55, 161 (1982).
6. Bolland, J. L., *Q. Rev. Chem. Soc.*, 3, 1 (1949).
7. Hawkins, W. L., and Worthington, M. A., *J. Polym. Sci.*, A1, 3489 (1963).
8. Kendall, F. H., and Mann, J., *J. Polym. Sci.*, 19, 503 (1956).
9. Layer, R. W., and Lattimer, R. P., *Rubber Chem. Technol. Rev.*, 63, 3, 426 (1990).
10. Erickson, F. R., et al., *Rubber Chem. Technol.*, 32, 1062 (1959).

11. Rozumouskie, S. D., and Batashova, L. S., *Rubber Chem. Technol., 43,* 1340 (1970).

12. Cox, W. L., *Rubber Chem. Technol., 32,* 364 (1979).

13. Andries, J. C., et al., *Rubber Chem. Technol., 52,* 823 (1979).

14. Ferris, S. W., Kurtz, S. S., and Sweely, J. A. *ASTM Technical Publication 229,* 1958, p. 72.

15. Jowett, F., The Protection of Rubber from Ozone Attack by Use of Petroleum Waxes, *Rubber World, 188,* 2 (983).

Miscellaneous Ingredients

I. INTRODUCTION

It is hard to imagine any mixture besides a rubber compound in which so many materials have been tried out. What other formulas, for example, would use coal dust, dried animal blood, cotton fiber, walnut shells, and cocoa butter? Yet coal dust has been used solely to get a cheap black compound, dried blood to improve adhesion in press rolls, cotton fibers to give stiffness in tubing, walnut shells for traction in winter tire treads, and cocoa butter to provide a more pleasant odor.

Obviously we cannot review here all the materials used in compounding not covered previously. However, some ingredients impart special properties and are needed often enough to justify at least some exposition of their characteristics. These include:

1. Blowing agents
2. Rubber chemical dispersions
3. Flame retardants
4. Colorants
5. Bonding agents
6. Mold release agents

II. BLOWING AGENTS

Blowing agents are used to make soft, light, impact-resistant sponge rubber. Most often used alone, but occasionally with other materials, they form gases at curing temperatures which put holes in the rubber and make it spongelike. If the open spaces or cells are interconnecting, the product is referred to as open cell sponge; if not, closed cell. Sponge rubber should not be confused with latex foams, which are made by whipping air into rubber latex and then vulcanizing, or chemical foams like urethane foams, where carbon dioxide is produced during the reaction of the isocyanate with water.

Sodium bicarbonate, the first blowing agent, reacts with stearic acid to give carbon dioxide as the temperature is increased. Because of the permeability of carbon dioxide, interconnecting cells are largely produced. Ratios like 1 part of stearic acid phr to 2 or 3 parts of sodium bicarbonate phr are used. A disadvantage of using sodium bicarbonate is that a residual soap is left in the sponge.

Closed-cell structures are made by using blowing agents that evolve gas, largely nitrogen, by thermal decomposition. Some of the common blowing agents in this class are *p*-toluenesulfonylsemicarbizide, azodicarbonamide, diazoaminobenzene, *N,N'*-dinitrosopentamethylenetetramine (DNPT), *P,P'*-oxybis (benzenesulfonylhydrazide) Although the latter has been marketed widely in Europe and the United States, it has the disadvantage of having a low decomposition temperature (so low that special precautions should be taken in its handling); it also leaves a unique, unpleasant fish odor in the product.

Again health concerns arise in using blowing agents, and one of the materials named above—diazoaminobenzene—is no longer marketed by one company because it releases unhealthy aniline fumes. Another point to be remembered in using blowing agents like these is that they leave considerable solid residues— over 50% of the original weight of the product.

The proper selection and use of a blowing agent are difficult tasks. Some characteristics that must be considered are as follows.

1. The blowing agent's chemical composition and whether activators are required.
2. Gas evolved per gram, tested by heating the sample in dioctyl phthalate in a flask, collecting the noncondensible gas evolved, and noting the gas evolution rate. A modified azodicarbonamide marketed by Uniroyal as Celogen 770 evolves 180 cm^3/g; DNPT will produce about 200 cm^3/g.
3. The decomposition temperature. This may vary from 100 to 235°C, but decomposition can occur at lower temperatures if catalysts are present, and there may be explosions. At the high end of the scale, few manufacturers have facilities that can cure at 230–240°C.
4. The decomposition products. Blowing agents do not completely decompose into gases such as N_2, CO_2, and CO; rather, they leave substantial residues in the sponge such as urazol and cyanmic acid, some of which may be objectionable.
5. Uniformly of color and cell structure produced in successive batches of stock.

6. Absence of odor.
7. Whether the blowing agent has an influence on vulcanization and aging—effects in both areas should be minimal.
8. Possible discoloration of the vulcanizate.
9. Cost.

Typical of the difficulties involved in making sponge are determining the right plasticity of the stock for proper blow (stocks that are too stiff hinder expansion) and synchronizing the rate of blow with the rate of vulcanization. In the case of natural rubber sponge, the raw rubber is peptized and brought to Mooney values in the range 15–30 before mixing. If the synchronization is not right and blowing occurs too rapidly, the expanded material may collapse before vulcanization is completed. Nitrogen blowing agents are usually used at dosages of 6 or fewer parts in the rubber. In open cell sponge the mold cavity is filled to only 30–70% of its volume. The air above the rubber in the mold can be bled off by having a fabric between the mold plate and the top plate. Because of the very fine honing of plasticity needed for good sponge making, as well as the necessity for excellent dispersion and little if any prior decomposition of the blowing agent, dispersions of the latter in a polymer binder are often used.

III. RUBBER CHEMICAL DISPERSIONS

One of the growth industries peripheral to rubber goods manufacturing during the last decade or so has been dispersion of rubber chemicals (usually provided by custom mixers). In the main, these are single chemicals dispersed in a polymeric binder, although occasionally combinations of chemicals are provided. Besides dispersions, powdered liquid concentrates and free-flowing powdered liquids (wetted powders) are available.

A wide range of rubber chemicals have been dispersed in polymeric binders and are vigorously marketed. The polymers used for the matrix are largely ethylene propylene rubbers (both the copolymer and the terpolymer), polyisobutylene, SBR, and NR. The dispersed material may constitute anywhere from 30 to 90% by weight of the dispersion. Some typical dispersions are as follows: 70 parts of the accelerator, TMTD, with 30 parts SBR, or 40 parts of dicumyl peroxide with 60 parts of polyisobutylene. These dispersions can often be found in pelleted form as well as in slabs. An example of a powdered liquid concentrate is one whose composition is 72 parts of a coumarone indene resin on an inert powder (28 parts), which makes it free-flowing. Compounders familiar with the trouble in measuring out, say, a coumarone indene resin with a 25–30°C melting point will appreciate that convenience. Wetted powders include such compositions as 90 parts of sulfur with 10 parts of a rubber processing oil.

Buying rubber chemicals in dispersed form is more costly than mixing from scratch. Incidentally, dispersion in plasticizers are somewhat cheaper than those using a polymeric binder, but usually the former have more extraneous matter and handle less easily.

There are solid advantages to the use of dispersions, powdered liquid concentrate, and wetted powders. Some of these are listed.

1. For hard-to-disperse materials, they facilitate the making of good dispersions.
2. They contribute to improved housekeeping in the plant. Compliance with government regulations on the disposal of scrap materials is easier if the materials are in rubber form.
3. Because of the improved shelf life and better dispersion, often less of the rubber chemical can be used with satisfactory results.
4. Mixing time and energy may be reduced (e.g., if a vulcanizing dispersion can be used at the batch-off mill from the Banbury rather than having another Banbury pass just to add the curatives).

A good illustration of the benefits that can occur from using a dispersion is given in a bulletin [1] by a company that manufacturers a dispersion of activated magnesia, 74 ± 2% plus 26 ± 2% organic binder for use in neoprene stocks. The latter have been historically cured with 5 parts of ZnO and 4 parts of lightly calcined MgO. The latter has two functions: to get as an acid acceptor and to inhibit scorch. Unfortunately light calcined magnesia is very hygroscopic. It picks up water from the atmosphere and tends to be converted to magnesium hydroxide. This destroys its function as a scorch retarder and makes processing difficult. Manufacturers of light calcined magnesium oxide protect their product by packing it in small-quantity polyethylene bags. This makes it more expensive and, of course, bags can tear or be prematurely opened. Using the dispersion alluded to, however (Scorch guard), scorch time was longer than with fresh Extra Light Calcined (ELC) magnesium oxide, even though the dispersed material had been exposed for a week. ELC is a grade of magnesia preferred for neoprene compounding. The batch using the dispersed material showed a slight improvement in tensile properties. Another advantage was that MgO in dispersed form could be reduced to about the 3 phr level with no reduction in stress–stain or scorch resistance properties. Finally, data are presented that show that mixing times can be dramatically decreased with resultant energy savings. This finding has been supported by factory trials.

Vulcanization of rubber extrusions in salt bath vulcanizers cannot be done if water present in the mix is not scavenged. At the temperatures used in this method, say 380°F, blisters and porosity would be caused by the escaping steam. There are a variety of moisture scavengers such as phosphorus pentoxide, anhydrous calcium sulfate, silica gel, and calcium oxide. Only the latter is practical and used. The other desiccants release water they have sequestered at about 150°C, a temperature that can easily be reached in extrusions. On the other hand, calcium hydroxide, formed by the reaction between water and lime, retains the water up to about 350°C, far above rubber processing temperatures.

Raw calcium oxide is rarely used as a desiccant; rather, it is used as a dispersion. Alone it is irritatingly dusty and hard to mix well, and it hydrates with moisture from the air. A typical product, DesiCal 85, marketed by Harwick Chemical Corporation, has 85% CaO with 15% of a high flash point rubber process aid as the binder.

Although stoichiometrically, 3 parts of calcium oxide react with 1 part of water, more desiccant is used in factory practice because of the difficulties encountered in getting an ultimate dispersion. Accordingly up to 10 parts of lime may be used per 100 parts of polyer.

IV. FLAME RETARDANTS

In the gamut of rubber products there are some that must be flame-resistant. Some examples are carpet backing and rug underlay, underground conveyor belts, rubber insulated wire and cables, and, more recently, some single-ply roofing materials. Some polymers such as neoprene are inherently flame retardant because the halogen released by the burning tends to quench the flame by keeping oxygen away. Other elastomers with flame resistance are chlorinated polyethylene, fluorelastomers, and silicones. Perhaps the most that can be done to improve the flame resistance of hydrocarbon polymers is to have a compound that does not continue to burn when the source of heat is removed.

Resistance to burning is enhanced, even with fire-resistant polymers, by the addition to the compound of flame retardants. In many cases mixtures are used, to take advantage of synergistic effects. Over the years a variety of additives have been used; the most useful division is between inorganic and organic agents.

Commonly used inorganics are clays, $CaCO_3$, $MgCO_3$, $Al(OH)_3$, $Mg(OH)_2$, Sb_2O_3, and zinc borate (variable composition). Sb_2O_3 functions somewhat differently from the others and is discussed later. The other products, when used at high loadings (e.g., 20% by weight in the compound), function by dilution of the polymer content, which is achieved by the conduction of heat away from the contact area with the flame. In the case of aluminum and magnesium hydroxides, flame temperatures are high enough to cause dehydration, with the adsorption of a considerable amount of heat, and steam is produced. Ammonium phosphates are also used as retardants. Like the hydroxides, they dehydrate, thus leading to forming polyphosphates with the adsorption of much heat inhibition of the flame propagation reactions.

By far the most common flame retardance system consists of two additives, and one of these is usually antimony oxide. The other is usually chlorinated paraffin, polychlorinated alicyclics, or brominated aromatics. The basic reaction is:

$$Sb_2O_3 + 6HCl \text{ (or HBr)} \rightarrow 2SbCl_3 \text{ (or SbBr}_3) + 3H_2O$$

The antimony trihalide is volatile, thus the gaseous atmosphere around the burning part is diluted, and steam is also produced.

Common organic flame retardants are chlorinated paraffins, polychlorinated alicyclics, and brominated aromatics. In these cases thermal decomposition releases HCl or HBr, which can inhibit flames by radical transfer with reactive species in the propagination reaction. A brief review of the mechanisms proposed has been given by Lawson [2]. Halogen-containing retardants are not as effective alone as with another retardant such as Sb_2O_3.

Some flame retardants in the United States are supplied as follows:

Material	Brand name	Marketer
Chlorinated paraffin	Chlorowax 70	Diamond Shamrock
Decambromodiphenyloxide	DE-83R	Great Lakes Chemical
Zinc borate	Firebrake AB	U.S. Borax
Phosphate ester	Kronitex	FMC Corp.

Because of the varied applications in which the flammability of rubber products might constitute a safety hazard, several bodies have developed flammability tests. These include the Upholstered Furniture Action Council (UFAC), the U.S. Department of Commerce, ASTM, the U.S. Bureau of Mines, and Underwriters' Laboratories. All these organizations are based in the United States. For European and other area testing institutions, refer to *International Plastics Flammability Handbook* [3].

There are few comprehensive studies on the flammability of commercial elastomers and their nonburning formulations. A most useful study, however, is that by Trexler [4]. No less than 28 flame retardants were investigated. Then the most efficient were used to formulate nonburning polymeric compounds, which were classed as either combustible or self-extinguishing.

Before leaving the subject of flame retardance it should be remembered that the materials are relatively expensive. In the amounts used to achieve high fire retardancy, moreover, they may alter or reduce wanted properties significantly. For example, a system might use as much as 10 parts of Sb_2O_3 and 30 parts of a chlorinated paraffin. Accordingly these materials are not used unless there is a definite requirement for flame-resistant properties.

V. COLORANTS

Although most rubber compounds are black, colored rubbers are frequently wanted for such consumer items as hot water bottles, household gloves, rubber toys, bathing caps, and, of course, white sidewall tires. Synthetic rubbers and natural rubber vary from a pale yellow waxy color to a dark brown, and pure gum articles have colors depending on the rubber source. Many such compounds are translucent, but as the zinc oxide dosage is increased they become opaque. Pure gum colors are more acceptable in footwear items, laboratory tubing, and bicycle tires.

When the color has been decided on, the properties of pigments to give that color are assessed against what is demanded in a rubber compound. Four areas of concern are chemical properties, physical properties, fastness properties, and cost.

Chemically, pigments for rubber coloration should not interact with the polymer or other ingredients. They should be free of impurities like certain manganese salts, which would catalyze oxidation of the rubber. They should be stable to the temperatures used in fabrication and vulcanization. Certain pigments, such as the chrome yellows, are losing favor because of real or suspected carcinogenicity.

Physically, the colorants should disperse easily and should be insoluble to the extent that they won't bleed to surface, for example, if the rubber compound has oil in it. A rubber colorant should have adequate covering power—a property often tested by dispersing the colorant in a liquid medium and seeing what concentration is needed to cover a black and white checkerboard design so that individual squares cannot be discerned. Formerly all colors were added in dry form. Now pigments in oil pastes or dispersed in polymers are preferred because they eliminate the flying around and loss of dry pigments.

Materials used to color rubber should have fastness with respect to light, solvents, and service conditions. Fastness to light is an obvious need for an article exposed to the sun; solvent resistance is required in, say, colored household goods used with cleaning solutions. Critical service conditions include use in golf club and tennis racquet grips, for example, where resistance to perspiration is called for.

Cost considerations are always a critical concern for compounders, and escalating costs have pushed some pigments out of reach for rubber articles in a competitive market.

Colorants for rubber articles can be classified as inorganic or organic—about the only broad classification possible, considering the array of materials available. We will return later to a discussion of inorganic and organic pigments and their use.

For simplification at the start, we can deal with black and white colored goods that use only inorganic pigments. Black rubber products are made using carbon black as the pigment. To get deep jet blacks, the finer particled, high tinting strength blacks like ISAF should be used. Jetter blacks are available, such as the carbon blacks used in the printing and protective coating industries, but in most cases their cost is prohibitive. No problem of fading, staining, or premature aging will occur with carbon black pigmentation. Titanium dioxide, zinc oxide, or lithopone can be used for white stocks. Because of its importance as a white pigment, titanium dioxide is reviewed in more detail.

Although titanium dioxide occurs naturally, only the manufactured product is used by the rubber industry. It is produced in two ways: by the chloride process and the sulfate process. In the former, titanium tetrachloride is burned in oxygen and yields titanium dioxide and chlorine. In the sulfate process an acid solution of titanium sulfate is hydrolyzed and the hydrous precipitate calcined. Titanium dioxide is produced in two crystalline forms: anastase and rutile.

Titanium dioxide is the predominant white pigment in the world for several reasons. It has high refractive index and stability, is nontoxic, and can be produced in the right size ranges.

Titanium dioxide is valued in rubber compounding for its high covering power and for giving the whitest whites. The rutile type is somewhat better than the anastase variety in this respect. 30–85 parts of rutile being equivalent to 100 parts of anastase. Titanium dioxide will assay 95% and up TiO_2, small amounts of Al_2O_3, and SiO_2 being present. Density is about 3.9 for the anastase, 4.1 for the rutile. At least in natural rubber mixes it appears to have no effect on vulcanization. Anastase TiO_2 has a blue cast and this, coupled with the characteristic brownish color of unpigmented NR or SBR stocks, gives a pure white.

Titanium dioxide is an expensive pigment, but the cost is partially compensated for by the greater brightness it gives per pound compared to other white pigments. For example, the equivalent of 100 parts of titanium dioxide would be over 550 parts of lithopone. Compounds using TiO_2 include many white compounds such as white sidewall strips, bathing caps, and athletic footwear like tennis shoes. Titanium dioxide is often blended with other white or light-colored materials like clays for these white stocks. For example, a bathing cap stock might contain 40 parts of whiting with 10 parts of titanium dioxide. In colored stocks it is often useful in preparing clear pastel shades with organic pigments.

In making a colored stock the usual procedure is to prepare a white stock and then add the required amount of colorant. One way of evaluating stocks with respect to the relative amount of color necessary is as follows.

Use a standard white stock as a masterbatch. This could be pale crepe 100, zinc oxide 3, blanc fixe 50, stearic acid 1.5, sulfur 2.25, MBT 0.20, and TMTS 0.35. If the same material is being evaluated from two suppliers, a small amount, say 1.0%, of material A is mixed with a specified portion of the masterbatch. The same proportions are used with material B. Both mixed samples are split, and an equal weight of masterbatch again is mixed with half of the original sample. We now have stocks containing 1.0 and 0.5% material A and others containing 1.0 and 0.5% of material B. The four samples can then be examined: two tensile slabs of each of the four stocks prepared but only one of each pair vulcanized. By examining the unvulcanized and vulcanized slabs, one learns the stability of the materials under curing, while by comparing the vulcanized samples the relative strength and shade of the colors tested and how they stand up to dilution can be ascertained.

Earlier it was mentioned that rubber pigments are broadly divided into two classes: inorganic and organic. Comparing the two, inorganic pigments are weak, often dull, and in some cases too opaque to impart the desired richness of shade. They are quite insoluble, do not bleed or bloom, and often have the required chemical resistance and heat stability. Usually they are fast to light, solvents, and service conditions. Poor dispersibility is often encountered with these pigments.

On the whole, less information is available about the organic colors. They generally give truer shades, are brighter, and perhaps 5–10 times stronger than the inorganics. On the other hand, they are more unstable and sensitive to heat and chemicals. On long-term exposure to sunlight they can fade badly. Above all, they are relatively expensive.

Rubber compounds should be designed for coloration. This means that staining antioxidants like the p-phenylenediamines should be avoided; phenolics can be used in their stead. Some curing systems may cause discoloration in curing. Peroxide curing generally shows little if any discoloration.

In formulating colored stocks, the compounder may have to come up with a dull color stock, possibly an earthy tone, a bright clean colored compound, or a pastel one. The first category would normally use inorganic colors with perhaps a low opacity whiting loaded stock. Some inorganic colors are:

Red: cadmium sulfide selenide, iron oxide
Yellow: cadmium sulfide, lead chromate, iron oxide
Green: chromium oxide
Blue: ultramarine (iron blue), cobalt blue
Brown: iron oxides

Pigments of these types have surface areas of 1–100 m^2/g. Iron blue is about 75m^2/g. There is always the question of the toxicity of any chemicals used in compounding. At the time of writing lead and zinc chromates were considered to be possibly carcinogenic, but cadmium pigments were thought to pose no health problems.

Oxides of iron provide a wide range of pigments from brown to red to yellow. They are of two kinds: natural and synthetic. The natural oxides are widely known artists' colors. The brown oxides are recognized as numbers or Van Dyke brown, the reds as red iron oxide, and the yellows as ochre. The synthetic counterparts of these—yellow, red, and brown colloidal pigments—have some advantages over the natural oxides They are chemically more pure, and they have more uniform particle size and size distribution. Specific gravities vary from about 3.90 to 5.20.

Cadmium pigments are brilliantly colored and offer a wide range from primrose yellow to orange-red and maroon. Golden yellow cadmium sulfide will have a specific gravity of about 4 78; mixed with lithopone, abut 4.55. Since cadmium pigments are expensive, forms that contain lithopone are often used.

If bright, clean colors or pastels are required, organic pigments probably should be chosen and used with a white base stock like one containing titanium dioxide. Some commonly used pigments are:

Red
  red pyrazoline
  Permanent Red 2B
  Lake Red C (orange-red)
Yellow: diarylide
Orange: diasnisidine
Blues and greens: phthalocyanine

Over the course of time, organic pigments have come to be identified by such terms as Permanent Red 2B or diarylide yellow AAA; reference to a paint chemists' handbook will help in identifying the particular pigment wanted.

Of the reds listed, pyrazolone is stable up to 350°F in molding, has fair light stability (fedeometer testing), resists bleeding into soap solutions, and can be used in open steam cures. Permanent Red 2B has slight bleeding in soap solution, can be cured up to 330°F, and has fair light stability. Lake Red C has poor light stability. The diarylide yellows have excellent heat and light stability, do not bleed in soap solution, and can be cured in open steam. The dianisidine organ has poor light stability and should not be cured at temperatures over 330°F. The blue and green phthalocyanine pigments show excellent heat and light stability, do not bleed in soap solution, and can be cured in open steam.

Organic pigments can be expensive. A good red pigment, for example, may cost $12/lb. For this reason the necessity for their use over inorganic pigments should be established and careful testing of the experimental compounds made. Tests might include fadeometer or weatherometer testing, outdoor exposure, testing for any migration or staining tendencies, and testing with solutions to see if there was any leaching. A homely example of this might be dish-draining pads, which are subject to intermittent exposure to detergent solution.

VI. BONDING AGENTS

Most large-volume rubber products are bonded composites—tires, hose, belts, footwear, and insulated wire. In certain cases such as hose and belting, bonding agents are not ordinarily needed. For example, by the use of cements, doughs, and thin calendered stocks, braided or wrapped hose can be made in which the physical locking of the elastomeric compound into the interstices of the fabric or yarn gives a sufficiently strong bond.

Bonding agents are used when physical adhesion is insufficient. Complicating matters is the variety of metals and fabrics (or yarns) to which the rubber must be bonded. The metals (usually mild steel, copper, brass, or aluminum) can be in sold metal plate or forms of wire. The textiles include cotton, nylon, polyester, rayon, and glass. In many cases cotton does not require a bonding agent.

Bonding agents include:

1. Ebonite
2. Brass plating
3. Chemical agents such as proprietary rubber-to-metal adhesives, isocyanates, resorcinol formaldehyde rubber resins, and postvulcanization bonding agents

If rubber-to-metal bonding is carelessly done, there may well be violations of environmental protection laws.

Choice of the bonding method depends upon the materials to be joined, their form, and the service conditions. If the bond is to be between a rubber compound and a metal, it is imperative that the metal part be clean. The first step in cleaning the metal part is to remove any oil, such as die lubricants. This is most often done by solvent degreasing, occasionally by alkaline cleaning. Once the oil has been removed, the next step is to rid the part of scale or oxide coatings that were insoluble in the solvents. This second cleaning is done by either mechanical or chemical means. Grit blasting is a typical mechanical method. Chemical methods include acid pickling, phosphatizing the surface with zinc or iron phosphate, and electroplating. Good protection of the cleaned parts from oil atmospheres, handling, and dust must be maintained if excellent bond strengths are to be realized.

Ebonite has not yet been described. It is a hard rubber that owes its hardness to sulfur vulcanization with approximately 30–35 parts of sulfur per hundred of rubber being used. Black hard rubber pocket combs are commonly seen ebonite products.

In ebonite bonding a thin layer of ebonite solution is applied to the metal part and then dried. This forms a layer to which the outer rubber compound can adhere, and the bond is formed when the part is vulcanized. Such a process might be used with the rubber-covered steel rolls used in paper making. The ebonite becomes thermoplastic at high temperatures, however, and the joint might fail from slippage. Again there may be poor aging between the ebonite and the cover layer as a result of the difference in sulfur and accelerator content. Occasionally a low sulfur tie gum is placed between the ebonite and the cover stock for this reason.

The brass plating method consists of electroplating the metal part with brass; the rubber compound reacts chemically with the brass during vulcanization to form a strong bond. This approach has been extensively used in adhering rubber compound to tank track blocks. Excellent bonds are possible by this method. There is, however, a continuous difficulty in keeping the plating solutions just right to prevent variation in the composition of the plate. Then too, the method is not suited for small-volume production or all rubbers.

Special bonding techniques are required to adhere rubber compounds to such synthetic fibers as glass, polyester, and nylon yarn or fabric. Here it is common to treat the fiber with a dip consisting of resorcinol, formaldehyde, and latex (RFL) and to have similar ingredients in the rubber for good adhesion. The inner body ply compound for a passenger tire would probably have resorcinol along with a methylene donor like hexamethylenetetramine to polymerize the resorcinol. Proprietary methylene donors (e.g., Mobay's Cohedur AS) frequently are used for such purposes. Adhesion is improved by the addition of a small amount, say 5–6 phr of precipitated fine silica.

An important bonding problem consists of obtaining good adherence of steel wire to its rubber insulation in radial steel belted tires, hose, and conveyor belts. Such tires can fail if there is severe loss of adhesion between the wire and rubber or if the wire itself becomes corroded. Water may be present in a tire from deep cuts in the tread area caused by broken glass, nails, etc. On reaching the wire insulation, this water can lower adhesion and corrode the steel wire at the cut ends. Steel wire is ordinarily plated with brass or zinc. Usually brass plating is considered to give adequate bonding to the rubber.

Because bonding strength can vary with the brass composition and plating thickness, higher bond strengths are sought by the use of direct bonding agents in the insulation stock. These bonding agents are usually cobalt soap compounds such as cobalt naphthenate or cobalt stearate. A recent product is Manobond C-16 from Manchem Incorporated, a complex metal organic compound based on cobalt and boron linked through oxygen to an organic component. Such agents are commonly used at a dosage equivalent to a cobalt metal content of 0.1 phr. Using these additives can result in a 35% increase in unaged adhesion. Because the insulating stocks are natural rubber, whose heat resistance is somewhat limited, bonding agents in the Manobond C series were tested after 10 days at 85°C and still gave 30–50% better adhesion than found with the control without additive. More details are given by the manufacturer [5].

Bonding agents that can achieve bonds with both textiles and metals are the isocyanate adhesives. These are usually proprietary solutions of organic isocyanates such as triphenylmethane trisocyanate in organic solvents such as xylene. One advantage they have is their flexibility; for example, mild steel, cast iron, aluminum, brass, and phosphor bronze have all been effectively bonded to rubber. Again they bond cotton, rayon, nylon, and polyesters to natural rubber and the principal synthetic rubbers. The isocyanate solution can be transferred to textiles by a dipping process or by spreading a rubber dough containing the agent over them. For metal bonding the isocyanate is coated on the piece by brushing, spraying, or dipping, and allowed to dry. If the rubber stock to be bonded is natural rubber or SBR, it is preferable to put a tie gum coat of rubber compound containing no sulfur or accelerator next to it, to prevent the reaction of certain groups, such as—NH_2 groups, with the isocyanates. Proprietary isocyanate bonding agents, for example, are Vulcabond TX marketed by Vulnex International and Desmodur R sold by Mobay. Along with the advantages of these materials come several disadvantages. They are, for example, very reactive, and any contact with moisture should be avoided. Health hazards are also involved, so effective protection of operatives is a must.

Finally we come to the bonding agents most often used for bonding rubber compounds to metals. These usually proprietary products are of undisclosed composition. Originally the base for these agents was chlorinated natural rubber. They are supplied as solutions/dispersions that can be sprayed, dipped, brushed, roll coated, or tumbled on the part. The chosen method depends on such factors as size and shape of the part, number of pieces, and whether partial or complete coating is required. Frequently these agents are applied in two steps—a prime coat followed by a top coat when the first is thoroughly dry. Common applications are providing the bond in automotive engine or body mounts and between the nipples and rubber components in oil suction and discharge and rotary drill hose.

These bonding agents are essentially solutions or dispersions of polymers or so-called crossbonding agents. Originally they were solutions in solvents with low flash points (say < 85°F) and a solids weight percent of 15–30; they were diluted with solvents like toluene, xylene, or methyl ethyl ketone. Such bonding agents were developed in the early 1940s, supplanting the brass plating technique in many bonding applications. These agents gave very satisfactory bonds but are now displaced somewhat because of environmental concerns. The newer agents, being water emulsions, eliminate solvent fumes. They also bring some unique difficulties: they can foam easily, they spoil if frozen, and they require distilled or deionized water for dilution. Two prominent suppliers of rubber-to-metal bonding agents in the United States are Lord Corporation (Chemlok adhesives) and Dayton Chemical Division (Whitaker Corporation) (Thixon adhesives). Both lines have water-based bonding agents: Chemlok 802 and 828 from Lord, Thixon 753 and 954 from Dayton Chemical Division (Whitaker Corporation).

There are few books devoted to the subject of rubber-to-metal bonding. There is much useful advice, however, ir a book published in 1959, namely the second edition of *Rubber to Metal Bonding*, by Buchan [6]. Compounders concerned with bonding should be familiar with the ASTM Standard on Adhesives, which shows detailed symptom description of bond failure. Also useful is a brochure by Lord Corporation [7]. Probably in few other rubber processing operations is the advice and counsel of the supplier as valuable as in the rubber-to-metal bonding field.

VII. MOLD RELEASE AGENTS

It may seem a broad extension of compounding technology to include mold release agents, but these materials are at times directly added to the compound, few molding processes do not use them, and they can cause production problems (e.g., surface disfiguring) if improperly used. In general, rubber compounds tend to stick to mold surfaces. Mold release agents (also called parting agents or mold lubricants) are used to prevent this sticking and thus facilitate part removal. They can also help to complete mold filling and improve the finish of the part.

Some materials—like metallic stearates—may be added to a compound as internal release agents, which function by migrating to the surface, helping to clear the part. Because internal release agents are rarely used, and often require an external agent as well, our attention will be centered on external agents.

Materials used as external mold release agents include:

Mica	Polyglycol oils
Talc	Metal stearates
Soap	Waxes/greases
Silicone oils	Fluorocarbons
Formulated compounds	

Except for the powders, and oils that ae used directly on the mold, these release agents are dispersed in liquid vehicles (water or solvent) and sprayed on the mold. There the water or solvent evaporates from the hot mold surface, leaving the release coating. Spraying can be from aerosol cans or commercial spraying equipment. Provided the latter is properly designed for the job, it is much more economical than aerosol spraying. Other ways to apply mold release agents include wiping, flow coating, painting, and roller coating. These usually are not as effective as spraying.

Sacrificial mold release agents are those that are applied to the mold before each molding cyle. When the cured part is removed, there is a a cohesive failure in the release agent, part of it being left on the molding, hence the name sacrificial. Some agents used in this way are animal and vegetable fats, soaps, polyvinyl alcohol and silicone oils. The silicone oils, usually polydimethylsiloxanes with molecular weights of 6000–8000, are used in the greatest volume, as solvent or water emulsions.

Silicone oil emulsions are on the whole good mold release agents, but they do have problems. For example, automotive companies do not want silicone oil in their plants and so discourage vendors from using mold release agents containing it. Some of the oil transfers to the part, but the remainder stays with the mold and over time builds up as a varnish, which is difficult to remove. That portion of the release agent remaining on the part can contribute to problems when a postvulcanization step like painting is involved, and the residual oil must be removed first.

Because of the difficulties associated with them, silicone oil based release agents have been supplemented with largely proprietary formula, thus avoiding many of these problems just described. Two typical compounds might be mentioned. Frekote 800 from the Dexter Corporation is described as a stable solution of highly reactive materials that form continuous microthin films of the active ingredient during application. As the carrier solvents evaporate, the polymer chemically bonds to the mold surface. This action leads to the formation of a crosslinked polymeric coating on the surface of the mold, giving multiple release without contaminating transfer for rubber molding. Its coverage is estimated at 800–1000 ft^2 per U.S. gallon. Another product, McLube 1700, marketed by McLube Division of McGee Industries, is described as a mold release agent that uses a tetrafluorethylene (TFE) polymer similar to Teflon but much lower in molecular weight (5–10 × 10^3 vs. 1–4 × 10^6 for a typical Teflon polymer).

Technology in the production of mold release agents is changing as environmental regulations increase. For example, trichlorofluorethane, a popular release agent diluent, is now being reduced if not eliminated because of its potential to destroy ozone in the stratosphere. Another regulation, which sets limits on volatile organic compounds (VOC), makes reductions in the organic solvent content mandatory. The obvious approach to compliance with such requirements is to use water-based products. Although these have disadvantages (e.g., sensitivity to freezing during transit), they have been developed as alternatives by many suppliers, including the two mentioned above. Rigby [8] and Finn [9] have contributed papers on the state of the art for these products.

REFERENCES

1. Wyrough and Loser Inc., Trenton, NJ, bulletin 50–2/977.
2. Lawson, D. F., Recent Developments in the Flammability of Elastomeric Materials, *Rubber Rev. 1986, Rubber Chem. Technol.*, **69**, 3 (1986)
3. Traitzech, J. *International Plastics Flammability Handbook*, English ed., Carl Hanser Verlag, Munich, 1983.

4. Trexler, H. E., The Formulation of Nonburning Elastomer Compounds, General Motors Research Laboratories, Warren, MI 48090.

5. Manchem Inc , Bulletin MANOBOND C-16, 105 College Road, East Princeton, NJ 08548.

6. Buchan, S., *Rubber to Metal Bonding*, 2nd. ed., London, Crosby Lockwood, 1959.

7. Lord Corporation bulletin Elastomer Bonding, A Material Selection, Application, Process, and Troubleshooting Guide, 2000 West Grandview Blvd., Erie PA 16514-0031.

8. Rigby, M., *Rubber World*, 204. No. 5 (1991).

9. Finn, G. A., Bulletin: External Release Coatings, McGee Industries, Inc., Aston, PA 19014.

16

Compound Development—Part 1

I. GENERAL PRINCIPLES

Having reviewed compounding materials, their functions and properties, we consider next how the information is used in solving the compounder's problems. A compounder has three areas of concern:

1. Present compound modification
2. Troubleshooting
3. Design of new compounds

Satisfactory factory compounds may have to be modified, often because of the unavailability of some ingredient. Perhaps a processing oil is now branded as possibly carcinogenic, and prudence dictates its replacement. Substitutions usually are not difficult to make. The original supplier often recommends an alternative; if not, a competitor probably will be able to. It then becomes a matter of comparing the old with the new in a laboratory trial and, if the results are satisfactory, having one or more tests in the factory with the new material. If these results are satisfactory, the new material is used henceforth.

Such factory experimentation is by its nature not always successful, and the compounder next hears the question, "What do we do with the rejected stock?" As a result of greater demand for statistical quality control by customers, as well as an increasing number of environmental restrictions, this question becomes more and more difficult to answer. The historical reply has been to work it away (i.e., blend it) in other regular factory stocks of similar type at 5–10% by weight. More fortunate is the plant in which a regular production of lower quality parts can accommodate a small dilution by a large variety of compounds. One factory with which the writer was associated obtained a large contract for cheap rubber tires for toys. Since the quality level demanded was minimal, an outlet for blends of experimental, improperly mixed, and off-quality stocks had thus been found.

Troubleshooting to determine the cause of failure of a rubber part in manufacture, or, worse, at the customer's plant, can be difficult. A separate volume could be written on the varied techniques and checklists that might be used. Indeed, in some cases the fault does not lie in the compounding. For example, humidity conditions can interfere with the drying of rubber cement applied to the braid on rubber hose, such that if the cover is applied to the hose while much solvent remains in the cement, adhesion frequently will be poor.

In troubleshooting, then, the first step is to try to determine whether the proper compound was used. In this connection, a procedure called miscompounding is helpful. One notes, in the course of the original development of the compound, properties obtained when certain key ingredients (e.g., zinc oxide) are omitted. If a troublesome batch later shows properties identical to those of a given miscompounded batch, the cause of the problem is obvious.

The characteristics of all factory compounds should be well noted. The tests may be simple determinations of color, density, and hardness, ranging through more sophisticated measurements such as volume swell and instrumented methods like thermogravimetric analysis. If the right compound was used, and if it had worked satisfactorily in the past, poor processing or improper use would seem probable and should be investigated next.

The most difficult compounding is developing a completely new compound. Steps in this process are as follows.

1. Determine the target, the qualities needed.
2. Design preliminary compounds, going through the functional materials needed, selecting specific ingredients, and determining how much of each will be used. Existing formulations and suppliers' literature can be helpful here.
3. Check the cost. If it is more than 20% away from the target figure, if available, redesign to a lower cost.
4. Laboratory-test proposed compounds.
5. If compounds fail to meet the quality target, redesign to correct deficiencies.
6. Continue the process until at least one qualifier is obtained.
7. Recheck cost.
8. If cost is within 10% of target, run a factory trial, and check processing behavior and vulcanizate properties under factory conditions.
9. If necessary make adjustments to meet technical and cost requirements.

10. When factory trials are successful, establish limits for the satisfactory compound and add to the compound formulary.

The first step in developing a new compound is to determine the conditions it must meet as specifically as possible. This may be easier if the buyer has a specification for the part or item. At the other end of the spectrum, the compounder may be asked to provide a stock for a completely new use—say, a part on a robot in a hostile atmosphere. The more information available, the more quickly a suitable compound can be developed. Questions to be asked might include:

1. Does compound have to meet a specification? If so, what are the requirements?
2. Are the specification requirements firm, or are deviations allowed?
3. If no specifications are available, can samples of a competitor's parts or product be obtained for examination?
4. What are service conditions for the product or part? Is it a gaseous liquid, or closed environment (e.g., a gasket)? What is the composition and temperature of the environment?
5. Is the compound in direct or indirect contact with food? If so, compliance of all ingredients with the latest FDA regulations is mandatory.
6. Is the rubber part under static and/or dynamic stress? If dynamic, what is the frequency and amplitude?

In rubber compound formulation recipes are built, as it were. A common order might be in six steps.

1. With the service environment carefully considered, the type of rubber (natural, neoprene, etc.) is selected. In the simplest decision, one polymer is outstanding for the job (e.g., NR for stationers' rubber bands). In other cases two elastomers might each provide adequate compounds, and the decision is based largely on cost. Snowmobile belts, for example, can be made from NR or SBR. Finally, service conditions may be unique and a blend is warranted, say when a small amount of NR is added to a CR rubber compound to get a controlled amount of swelling in a product in contact with oil. With the selection of the polymer the further decision on the best grade of that rubber for the product.

2. The second choice is for the most appropriate filler and the amount. This of course does not occur with pure gum compounds. The color desired, the hardness (if specified), and the service environment will be some of the factors in that decision.

3. Next on the list is the vulcanizing agent; between sulfur, sulfur donors, metallic oxides (for neoprene), peroxides, urethane crosslinkers, or resin cures in the case of butyl. This decision is somewhat easier than the others since it depends largely on the type of polymer and the vulcanizing capabilities and practice in the plant. GN-type neoprenes regularly use metallic oxides as vulcanizing

agents; urethane crosslinking agents would be improper if open steam curing was used.

4. With the polymer, filler, and vulcanizing agent known, the accelerators and activators are selected, primarily on the basis of the vulcanizing agent chosen; next the polymer, and then curing and service conditions. A peroxide cure, for example, will not use accelerators or activators save zinc oxide. A GN-type neoprene uses metallic oxides as curing agents and rarely uses organic accelerators, staining accelerators cannot be used in light-colored goods.

5. Next are the plasticizers and/or softeners, which must be compatible with the elastomer and effective with the type of filler; they must not, of course, cause problems of their own. Paraffinic oils are incompatible with neoprene, high melting point hydrocarbon resins would be inappropriate in an NR/ISAF black stock (too stiff), and aromatic rubber process oils would discolor white or light-colored stocks badly.

6. The last fundamental question to be resolved is the age resister package. Two conditions must be satisfied here: providing suitable protection against the environment, and not choosing agents inimical to the softeners or the curative system. For example, antiozonants should not be used in a peroxide-cured compound; rather, one should use wax and an antioxidant.

These questions must be answered with almost every compound except pure gum compounds. Obviously for some compounds further choices are necessary: fire retardance, electrical conductance, or a particular color hue may be wanted, and these requirements must be filled with other materials or by replacement of one of the materials already selected. This might mean replacing part of the plasticizer with chlorinated paraffin to get flame retardance.

Before showing how a new formula might be built up, some generalizations can be made toward meeting specific requirements for selected properties. We can offer only generalizations, since hard and fast rules are comparatively rare in rubber compounding. Before going further, one point might be made. Good factory compounds tend to be lean on the number of ingredients. With the range of materials now available, a laboratory compound with 18 ingredients in one supplier's literature seems out of date.

Minimum tensile strength may still appear on some specifications, even though its value as a quality characteristic has been discounted. Tensile strength is obtained by using crystallizable rubbers like NR and CR, the finest particled black (silica in light-colored stocks), resinous rather than liquid softeners, and higher than normal sulfur if that is the vulcanizing agent. Not quite matched in this regard with oil furnace blacks, earlier blacks like easy and medium processing channel would give tensile strengths as high as 5000 psi with natural rubber. Overcured compounds lose tensile strength.

Occasionally modulus ranges must be met. In regular black-loaded compounds, 300% modulus for the same loading would vary widely depending on the polymer. In terms of descending modulus, highest would probably be CR followed by SBR, NR, IIR, and BR. EPDM is not included because it is usually a heavily loaded stock and cannot logically be compared with the others. Large modulus differences are obtained by the choice and amount of the fillers used. Reference to Chapter 10 will show what differences blacks can make. Nonblack fillers yield high modulus values with silica and the silicates, low values with such materials as ground whiting.

If there is a requirement for elongation, it usually is a minimum value. Since this property is almost the converse of modulus, compounding for higher elongation uses techniques that are the reverse of those used for increasing modulus. Lower loadings of the finer particled fillers would be appropriate, along with the use of liquid plasticizers and softeners rather than viscous liquids or hard resins. Undercures show higher elongations if that state of cure is permitted.

Other considerations aside, low hardness stocks can be obtained by more mastication of the rubber, if the material is amenable to breakdown, by using a minimum of fillers and a maximum of softeners and plasticizers. Softeners like rubber processing oils, factice, and lanolin are helpful. Hardness can be increased by using such materials as shellac, glue, fine particle silica, phenolic resins, and ground coal. For example, shoe soling can be hardened considerably by the use of themosetting phenolic resins. The hardness values reached (Shore A 90-100) could not be obtained with loading alone because the accompanying processing is very difficult. If a natural rubber or SBR stock is involved, significant hardness increases can be obtained by increasing the sulfur.

It is somewhat difficult to make recommendations on compounding for increased abrasion resistance. The latter can refer to how well a sandblast hose tube, a shoe sole, a conveyor belt cover, or a tire tread performs. All four products must be able to resist abrasion, but the service environments differ widely. For example, good sandblast hose tubes have been made with a pure gum natural rubber compound. The same compound would give very limited wear as a tire tread. Another complication is that abrasion resistance of a polymer may vary with the temperature. Winter tire tests with which the author was associated showed, for example, that in a northern city like Montreal in Canada natural rubber treaded tires had considerably less wear than the conventional SBR-treaded tires during the winter months (December through March). This tread wear relationship was reversed with warm weather driving.

For most abrasive wear situations, such as shoe soling, conveyor belts carrying crushed stone or ore, and tires, the rubbers of choice are NR, SBR, BR, or mixtures of the three. For U.S. passenger treads, blends like 70 parts of SBR/30 parts of BR might be used. Abrasion resistance in tire tread stocks is increased as finer particled blacks are used; however, as particle size decrease good dispersion is more difficult to obtain. Since good dispersion is essential for good wear, there must be tradeoffs. Abrasion resistance is often improved in such items as off-the-road tires by adding fine particle silica with the black. In white stocks, fine particle silica gives the best wear.

Compounding for solvent resistance depends on what solvent is used and the conditions of exposure. Only petroleum oils and petroleum-derived solvents are considered here. Reference to the literature should indicate the best polymers for other solvents.

Of the polymers covered in this text, three are considered to be oil-resistant: neoprene, nitrile, and polysulfide rubber. As noted earlier, petroleum oils can be classified as paraffinic, naphthenic, and aromatic in composition. Aromatic oils possess the greatest swelling effect; paraffinic the least. If high resistance to aromatic oils is needed, neoprene probably would be unsuitable and a butadiene-acrylonitrile rubber would be a first choice. If even that did not give sufficient resistance, a probable next choice would be a polysulfide rubber. Applications requiring such resistance might te paint spray hose tubes or the tubes of oil suction and discharge hose used to convey benzene.

Outside of polymer choice, there are certain ways to limit a compound swelling and thus usually increase its utility in service. One is to increase the loading. The percentage a compound swells is directly related to the percentage by volume of a polymer in a compound thus higher loadings mean less swell. Apart from this, there appear to be differences in swelling at the same loading of different fillers. For example, at the same loading MT stocks appear to swell more than SAF stocks in natural rubber; HAF are stocks intermediate. Another expedient is to preswell the rubber by including the same oil in the formulation that the product will contact or be tested by. Finally swelling is hindered by having tightly cured stocks.

Compression set is an important requirement in many applications, and low values are desirable. The test method most often used is ASTM method B, 22 hr at 70°C (D395). It is hard to reconcile some of the values in the literature; the test may have more inherent variability than most. Based on polymers alone, highest compression set values go with NR and SBR; CR is intermediate, and rubbers like IIR and EPDM give the least. Choice of black does not appear to have much effect except in neoprene, where thermal blacks give lower compression set than fine particled blacks like ISAF. While fillers cause higher compression set values than blacks at equal loadings, especially clays. There are two ways to reduce compression set by the choice of curing system. Efficient vulcanizing systems, discussed earlier, can reduce compression set perhaps 25–50% when they replace conventional systems. Alternatively, if other conditions permit, peroxide-cured polymers show excellent low set values.

The paragraphs above describe a few of the stratagems compounders use to meet the service demands on their products. Before going on to show typical steps by which some compounds might be built up, we return to one of the points touched on in the introduction, namely the compounder's responsibility to use a myriad of ingredients in a safe way to produce rubber compounds that carry no health hazards.

A compounder's activities involve health and safety concerns in three general areas: warehousing and laboratory storage of the ingredients, laboratory environment conditions, and characteristics of the rubber products formulated. All these areas are extensively regulated by the U.S. government, and many foreign countries have similar restrictions. The compounder must be aware of these rulings by having up-to-date copies of them. It is also important, in the United States, to follow the recommendations of such bodies as the Rubber Manufacturers Association or the Manufacturing Chemists Association, and—probably the most important—to keep in regular contact with suppliers, who are abreast of the changes affecting their products.

Simple examples of these problem areas include the proper storage of explosive materials such as the peroxides and the handling in the laboratory of fumes from solvents such as benzene. As another example, consider the need to be sure that one is not using, say, 15% by weight of oil furnace black in a hose tube stock for carrying cottonseed oil. (The writer can see no compelling reason for such a restriction, but U.S. law contains one.)

It is now a rare compounding ingredient that does not have supplied with it a Material Safety Data Sheet. A typical MSDS includes the following information:

Identification
Special regulatory hazards
Physical data
Fire and explosion hazard data
Reactivity data
Special protection information
Storage, spills, and disposal information
Health-related data

Every compounder should have a file of these data sheets on each of the materials used in his plant; a typical MSDS is given in Figure 16.1. Of the three areas mentioned above, two are covered by two federal regulations: the Occupational Safety and Health Act of 1970 (commonly called OSHA) and the Resource Conservation and Recovery Act of 1976. Seeing that compliance is in effect (e.g., proper storage of dicumyl peroxide) is not usually the responsibility of the compounder but rather the safety department of the company. However, the compounder is in a unique position to help compliance. The Resource Conservation and Recovery Act impinges directly on the compounder's work as regards scrap. Strict rules deal with its disposal. One example might be given of how far these concerns reach. Pallets of natural rubber imported into this country routinely have the wood treated to prevent wood borers from propagating. In the

MATERIAL SAFETY DATA SHEET

TRADE NAME: **AKROSPERSE 660 RED MB**

CHEMICAL NAME: Dispersion of Pigment Red 48:2 in SBR

CHEMICAL FAMILY: Mixture   CAS # 7023-61-2 (Pigment)

PREPARED BY: T. L. Miller   DATE: 07/20/92

••••••••••••••••••••••••SECTION II—HAZARDOUS COMPONENTS ••••••••••••••••••••••••

                                        OSHA Pel   ACGIH TLV   OTHER

None

Contains no SARA Title III, Section 313 notification chemical present at or above the de minimus concentration.

[Ingredients not precisely identified are nonhazardous. All ingredients appear on the EPA TSCA Inventory.]

•••••••••••••••••••••SECTION III—PHYSICAL/CHEMICAL PROPERTIES•••••••••••••••••••••

Boiling point: N/A   Specific gravity: 1.18 calc

Vapor pressure (mmHg): N/A   Melting point: N/A

Vapor density (air = 1): N/A   Evaporation rate: N/A

Solubility in water: insoluble

Appearance and odor: red rubber strips; little odor.

•••••••••••••••••••••••••SECTION IV—FIRE AND EXPLOSION DATA•••••••••••••••••••••••

Flash point: N/A   Flammable Limits: LEL: N/A   UEL: N/A

Extinguishing media: Water, foam, CO_2

Special fire fighting procedures:
  Wear self-contained breathing apparatus

Unusual fire and explosion hazards: None known

•••

Information presented hereon has been compiled from sources considered to be dependable and is accurate and reliable to the best of our knowledge and belief, but is not guaranteed to be so. Since conditions of use are beyond our control, we make no warranties, expressed or implied. In addition, if this document is reproduced, it should be done so in its entirety.

•••

Figure 16.1 Typical material safety data sheet (Courtesy of Akrochem Corp.)

MATERIAL SAFETY DATA SHEET (*Continued*)

AKROSPERSE 600 RED MB   Page 2 of 2

••••••••••••••••••••••••••••••SECTION V—REACTIVITY DATA••••••••••••••••••••••••••••••••••

Stability: Stable __X__   Conditions to avoid: None known
             Unstable _____

Incompatibility: None known

Hazardous decomposition products: None known

Hazardous polymerization: Will occur _____   Will not occur __X__
        Conditions to avoid: None known

••••••••••••••••••••••••••••• SECTION VI—HEALTH HAZARD DATA •••••••••••••••••••••••••••••

Routes of entry: Inhalation? No   Skin? No   Ingestion? No

Health hazards (acute and chronic): None known. Polymer-bound dispersions preclude the possibility of airborne dust. They also eliminate the problems generally associated with powdered chemicals.

Carcinogenicity: NTP? No   IARC Monograph? No   OSHA Regulated? No

Signs of symptoms of exposure: None known

Medical conditions generally aggravated by exposure: None known

Emergency First Aid Procedures:
  SKIN: Wash thoroughly with soap and water

•••••••••••••SECTION VII—PRECAUTIONS FOR SAFE HANDLING AND USE •••••••••••••

Steps to be taken if material is released or spilled: Pick up and return clean material to container for use. Place contaminated material in appropriately marked container for disposal.

Waste disposal method: Landfill or incineration, in accordance with federal, state or local regulations.

Precautions to be taken in handling and storage: Store in a cool, dry area

••••••••••••••••••••••••••••• SECTION VIII—CONTROL MEASURES •••••••••••••••••••••••••••••

Respiration protection: None required under normal conditions

Ventilation: Local exhaust: (desirable)   Special _____
              Mechanical (General) _____   Other _____

Protective gloves: Recommended   Eye protection: Safety glasses

Other protective clothing or equipment: None required

Other Precautions: None required

HMIS Rating: Health: 0   Flammability: 1   Reactivity: 0

consuming plant the wooden pallets are considered to be scrap, but firms are somewhat reluctant to give them away as firewood due to possible hazardous fumes being produced. More common are the problems involving the disposal of scrap rubber and compounding materials. These concerns are addressed in more detail in Chapter 20. Since the regulations are more liberal when the hazardous materials are mixed with elastomers, dispersions are often used. Diethyl thiourea often is purchased in this form.

Apart from ambient atmosphere and scrap regulations, the compounder must be aware of restrictions placed on compounds that may be in contact with food products. In the United States these are given in the Code of Federal Regulations 21, parts 170–199, revised every April. Section 173.2600 covers rubber articles intended for repeated use. As of April 1991 some of the restrictions were:

1. Accelerators—total not to exceed 10% by weight of the rubber product.
2. Plasticizers—not to exceed 30% by weight of the rubber product.
3. Fillers—carbon black—not to exceed 50% by weight of rubber compound, furnace combustion black content not to exceed 10% by weight of rubber products intended for use n contact with milk or edible oils.

Each classification by function is followed by a list of materials to which the restrictions apply.

More recently, other regulations have been appearing which can limit a compounder's flexibility. States and cities are enacting "right-to-know" laws regarding chemicals used in their jurisdictions. Essentially, chemicals the lawmakers consider to be hazardous must be labeled, and information about their composition and characteristics must be made freely available to the workers where they are used. Although the purpose of the laws is understandable, two side effects may be disadvantageous. Despite precautions, it becomes more difficult to maintain trade secrets. Also laws may foster an atmosphere conducive to increases in unwarranted grievances. For example, grievances might occur over the odor emanating from curing of a particular compound. Even if no health hazard were involved, the compounder might be under some pressure to change the compound to avoid a grievance.

As a reference on the toxicity of chemicals used in a rubber plant, U.S. compounders should find *Dangerous Properties of Industrial Materials* useful. Written by N. Irving Sax, published by Van Nostrand Reinhold, it is now in its 7th edition—a tribute to its value. Overseas readers might consider two other references: *Handbook of Reactive Chemical Hazards*, by L. Bretherick (Butterworths, London, 1985), and *Toxic and Hazardous Chemicals Safety Manual* (International Technical Information Institute, Tokyo, 1985). The latter appears to give more references to rubber chemicals.

II. ILLUSTRATIVE EXAMPLES

As a first illustration in building a compound, consider the formula that would be required for a conveyor belt cover. The belt is to be used primarily to carry ore or gravel in open pit mining in the western United States. Specifications for belts used in this particular service are lacking, and it is assumed that samples of a competitive belt considered to be satisfactory are unavailable. We can simplify the problem somewhat by considering the stress resisters in the belt to be plies of belting duck rather than steel cables. The latter case calls for special care in compounding, to ensure that the cover cures, as it were, in harmony with the insulation stock around the wire, which requires excellent metal-to-rubber adhesion.

We know the general requirements for this belt and its cover. The cover must be an abrasion-resistant compound to resist the pounding of the ore and the hitting of the cover by sharp edges, which tends to cause cuts and tears. The belt is exposed to the weather at all times, hence must be able to resist the abundant sunshine and high ozone concentrations occasionally in that area.

To make the belt endless when installed, or to permit repair of a torn section, it must be spliceable. That means stripping the cover from an end section of the belt, then successively removing shorter lengths of the exposed plies. The same thing is done in reverse at the other end of the belt. The two ends can be treated with an adhesive, mated together, and recurred in a portable press, to ensure that the spliced portion is not reduced in strength. The cover stock, as well as all other rubber compounds in a belt that will receive a second cure, must be resistant to overcure. If the belt were badly overcured, the rubber might soften and become unusable.

We start by selecting the polymer. The major requirement of the cover stock is high abrasion resistance. A review of the properties of the elastomers suggests natural rubber, SBR, and polybutadiene as logical choices. Because of the difficulty of processing all polybutadiene stocks, mentioned earlier, we discard BR rubbers. This leaves NR and SBR. Both can make good conveyor belt covers. However, a careful review of the literature indicates that natural rubber compounds have better tear resistance—a property of value here. Furthermore, at the time of writing natural rubber is cheaper.

Tentatively, we choose natural rubber. Remembering that polybutadiene has excellent wear resistance too and is low in price, we somewhat arbitrarily choose a blend, 80 parts NR and 20 parts polybutadiene. Reviewing the rubbers available, we settle on a technically specified grade, SMR 10. This uniform rubber has a limit on dirt content that ribbed smoked sheet rubbers do not have. High dirt levels have been know to accelerate failure from fatigue for rubber parts. For the best abrasion, a high cis polybutadiene should be picked such as CIS-4. So the formula begins:

| Natural rubber SMR 10 | 80 |
| Polybutadiene CIS-4 | 20 |

The next choice is the filler or reinforcer. In this case utmost reinforcement is wanted; there is obviously no need for a white compound, so carbon black will be used. As noted earlier, abrasion resistance increases as particle size decreases, so we consider only blacks in the HAF-to-SAF range. Because of its difficult dispersion and higher costs, we eliminate SAF. Both the remaining blacks, HAF and ISAF, could be used. We choose ISAF because it is somewhat more abrasion-resistant in most laboratory and road wear tests. In natural rubber both HAF and ISAF blacks appear to show peak wear resistance around 45–50 parts loading; we select 45 parts, the stiffer black, since we want fair processing.

The next question is what curative system should be used. A review of Chapter 7 makes it obvious that the less common cure systems that might be used with this polymer blend bring no special advantages. So the idea of using a peroxide cure or a urethane crossl nker is discarded. This leaves a sulfur curing system with the question of whether we should use a conventional sulfur cure or the semiefficient or efficient cures The last, it may be remembered, use sulfur donors instead of elemental sulfur. Any of the three could be used here. Although the efficient systems give good heat resistance, we do not need a great deal of heat resistance in this service, so we choose a conventional sulfur and accelerator system that is cheaper than the efficient system. Since the blend is 80% NR, we will build the formula as if it were 100% NR. The recipe can be fine-tuned later when the first laboratory results are obtained. Sulfur at 2.5 parts is within the conventional range, and we will use that. In choosing accelerators, we must remember that because of the requirement for spliceability, the cover stock must be resistant to overcure. To avoid precure in mixing and to ensure overcure resistance, a sulfenamide accelerator is chosen: n-oxydiethylenebenzothiazole-2-sulfenamide at 1.4 parts. As a secondary accelerator, TMTD might be used at 0.2 part. Since this is a black stock we do not have to worry about discoloration.

The formula now needs a plasticizer or softener. It this case we need neither resin-type plasticizer to maintain tensile strength nor ester-type plasticizer to provide flexibility in extreme cold. What is needed is a softener to make mixing a little easier and to help disperse the black. Thus a naphthenic rubber-processing oil, which has a good balance of processing properties for this blend, is selected.

For activation there is no need to depart from the standard stearic acid–zinc oxide combination. We choose 2.5 parts of stearic acid and 3.0 parts of zinc oxide. As discussed earlier, practice some time ago was to use 5 parts of zinc oxide, but more research has indicated that little is gained by going past 3 parts.

Finally, we come to the system to protect against aging. We need a really effective system in this environment, which has high temperatures in the summer (100°F not uncommon), some daylight ozone concentrations that might reach

10–20 per 10^8 parts of air, and intermittent operation of the belt. To obtain sufficient heat resistance at these ambient temperatures, we would use 2 parts of an antioxidant like polymerized 2,2,4-trimethyl-1,2-dihydroquinoline. This antioxidant has a low volatility and so will remain in the compound for some time. We need ozone protection, too, and one of the strongest protectors would be one of the alkylaryl *p*-phenylenediamine types; 1.5 parts is a common dosage. We can supplement the action of the antiozonant by adding 1.5 parts of paraffin wax. When the belt is not operating, the wax will form a protective film around it.

The complete formulation now is:

Natural rubber SMR 10	80.0
Polybutadiene CIS-4	20.0
Stearic acid	2.5
Zinc oxide	3.0
Rubber process oil	4.0
ISAF black N220	45.0
Antioxidant	2.0
Antiozonant	1.5
Paraffin wax	1.5
Sulfur	2.5
Amax*	1.4
TMTD	0.2
	163.6

*Vanderbilt's brand of *n*-oxidiethylenebenzothiazole-2-sulfenamide.

The density of this stock in the uncured state is 1.11, as determined by the method given in Appendix 1. Since there is no contact with food here, compounding ingredient choices can be freer.

There are 12 ingredients in this compound, each fulfilling a specific purpose. Possibly a peptizer should be added to facilitate breakdown of the natural rubber, but the stock could be mixed either way. When this stock is cured, typical vulcanizate values might be:

300% modulus	1550 psi
Tensile strength	4250 psi
Elongation	510%
Shore A hardness	63

As a second example of formula building, consider a hose tube suitable for conveying vegetable oils—for example, cottonseed oil or olive oil. Here the service conditions demand a stock that will have very low swell in carrying fluids (in the worst case the tube would swell shut), is extrudable, and will not impart any taste, odor, or color to liquids passing through. Other requirements include at least a moderate amount of heat resistance (the oils might be pumped warm) and, since these are edible oils, the compound must meet FDA requirements.

The literature on the polymers indicates that neoprene is highly resistant to fluids of the foregoing description Because such resistance is a prime requirement, neoprene will be used. Neoprene W is selected over neoprene GN because the former has better heat resistance.

We have some limited options in the case of fillers. For example, carbon black would be the first choice for the filler, but FDA regulations allow only 10% by weight of the compound to be oil furnace black, a kind we ordinarily would choose. In the interest of economy, we could load the compound heavily with fillers and softeners. Since, however, edible oils at elevated temperatures might extract significant quantities of the softener, we settle on largely white fillers with minimum softener content. For flexibility of the hose, a hardness of 60 ± 5 Shore A hardness units would be appropriate for the tube. With these points in consideration, we settle on 45 parts of a soft clay and come close to the 10% limitation by using 15 parts of an SRF black with it.

The next step is determining the cure system. Besides the standard 4 parts of light-calcined MgO and 5 parts of zinc oxide, neoprene W requires an organic accelerator. Ethylene thiourea is recommended by the neoprene supplier, so we add it to the compound at the 0.5 part level.

Some softener should be added to the compound to prevent it from sticking to the mill, to wet the pigments, and to help ease extrusion. We will add stearic acid, to minimize sticking, as recommended, but only up to 0.5 part, since it does retard cure. For the remaining softener a petroleum hydro-carbon resin like NEVCHEM might be used. This resin, made by the Neville Chemical Company, is an approved material by FDA standards. With a softening point of 72°C (ring and ball method), it should speed up extrusion yet at lower temperatures keep the uncured tube from sagging; 5 parts should be ample.

The next function to consider is age resistance. A hose tube such as this is not exposed to sunlight or to high concentrations of ozone, so no antiozonant is necessary. About the only protective agent needed is an antioxidant to resist deterioration by heat. Again an FDA-approved material is necessary, and one that will not stain or discolor. Such a material is 4,4'-thiobis-(6-*tert*-butyl-*m*-cresol). This agent is marketed by Monsanto as Santowhite Crystals. A usual dosage is 1.5 parts. The full formulation is now:

Neoprene W	100.0
SRF black	15.0
Soft clay	45.0
Magnesium oxide	
(light-calcined)	4.0
Zinc oxide	5.0
Ethylene thiourea	0.5

Stearic acid	0.5
Hydrocarbon resin	
(Nevchem 70)	5.0
Antioxidant	
(Santowhite Crystals)	1.5
	176.5

The density of this compound would be 1.52; cured 20 min at 307°F, it should show approximately the following properties:

400% modulus	850 psi
Tensile strength	2700 psi
Elongation	600%
Shore A hardness	62
Compression set	18–20%
(ASTM method B: 22 hr at 158°F)	

Volume increase after 1 week at 77°F would be 5% for both cottonseed oil and olive oil. Most technologists would consider such volume swells satisfactory for this kind of service. Although the swelling values were determined at only slightly above room temperature, it is unlikely that the oil would be pumped at, say, 200°F, where the volume swell would be significantly greater but the chance of taste impairment considerably increased.

As a final example, consider the formulation of an extrusion compound. Although we do not have a buyer specification, we are told that it is for an auto-motive item (accordingly very competitively priced), is black, and has no oil resistance requirement; good weathering resistance is wanted, however, includ-ing resistance to ozone. (There are relatively high ozone concentrations at times under the hood). We are further told of two requirements common to automotive specifications: hardness and tensile strength. The hardness must be 60 ± 5 Shore A units, and the tensile strength must be at least 1500 psi.

Reviewing our polymers, we can eliminate neoprene, nitrile, and polysulfide rubber, since oil resistance is not required and these compounds are relatively expensive. It is doubtful that an all-reclaim stock could give the tensile strength required. This first screening leaves NR, IIR, SBR, EPDM, and BR. Polybuta-diene is normally not used alone and would require high black and high oil loadings to be cost-competitive. Although both NR and SBR can be made ozone-resistant, antiozonants are expensive at the levels needed, so we discard these rubbers. This is done with some reluctance because there are some high black, high oil SBR masterbatches ordinarily used for retreading that might be useful here. Butyl has excellent weathering resistance, but with heavy loadings of black and oil we might not have enough margin over the minimum tensile requirement. This leaves EPDM—it can be extended a good deal and yet retain good rubbery properties. It also has excellent weathering and ozone resistance, so we will use this polymer.

Picking the polymer is easier than picking the right type of EPDM, with some 50 types offered in the United States by domestic manufacturers. For simplicity in this first trial we'll choose an unextended rubber with a nominal Mooney of 60 at 100°C. Most grades of EPDM use ethylidene norbornene (ENB) as the diene, which gives somewhat faster cures than, say, EPDMs with dicyclopentadiene. This should be an advantage. So still a bit arbitrarily, we choose Uniroyal's Royalene 501, which has a nominal Mooney of 60 at 100°C, uses ENB, and is recommended for mechanical products such as this.

Since the compound is to be black and there is a tensile strength requirement, we will only use blacks as filler. At this time we would ordinarily work out the amount and type of the black, then proceed to the curative system and the softener choice. Here we diverge from that procedure. We must keep the cost down, and since this is a question of how much black and oil we can add, we decide these amounts first. EPDM can be used with 50 to over 400 parts of black, along with increasing amounts of oil. We have to start somewhere, so we choose 150 parts of black. We want good extrudability, which is helped by high structure blacks, yet such blacks increase hardness. So we split the black, loading 120 parts of a high structure GPF along with 30 parts of SRF black—minimal structure but effecting less hardening of the compound.

The next step is to determine the hardness of the compound if there is no oil. Again we go to the literature, specifically focusing on loading studies of blacks in various polymers. If the hardness of blacks at various loadings in non-oil-containing EPDMs is extrapolated to zero loading, what might be called a base hardness for EPDM will be determined at about 50. Further studies of such tables indicate that a GPF HS will increase hardness by about one unit for every 2.2 parts of that black, but that 2.9 parts of SRF will be needed for the same unit increase. So hardness without oil would be 114.89 (50.0 + 120/2.2 + 30/2.9). This is above the Shore A scale and Shore hardness probably can't read closer than 1 unit, but rationality can be brought in at the end of the calculation.

Since the desired hardness is 60, we need to reduce the hardness by 54.89 units. It is known that for EPDM polymers, about 2.5 parts of oil is needed to decrease hardness one Shore unit. So we will add 54.89 × 2.5 = 137.2, rounded off to 137, parts of oil. Naphthenic oils are preferred for EPDM, so we limit our selection to those. Since this is a black compound, we are not limited by oil color or staining. A suitable oil might be Sun Oil's Circosol 4240, a high viscosity oil that might help avoid sagging of the compound in the uncured state. Another advantage is its low volatility, which should help heat resistance. Although not demanded for this application, heat resistance is frequently sought in automotive specifications and might give our compound a slight edge.

Activators are required, and these are regularly zinc oxide and stearic acid. As discussed earlier, zinc oxide functions well at 3 parts so we will use that, and stearic acid at 2 parts should be sufficient. Normally the next choice would be the protective system against aging. EPDM has such good resistance to weathering and ozone that protective agents of this sort are not needed.

The final step is the curative system. Many curative systems are available for EPDM. Although we want processing safety, the best systems for that rely heavily on low sulfur, high accelerator dosages, which tend to bloom. To avoid blooming, we pick a simple conventional system that can modified later if required, namely sulfur 1.50, TMTD 1.80, and MBT 0.60. The complete formulation is as follows:

EDPM Royalene 501	100.0
GPF HS black N650	120.0
SRF black N774	30.0
Circosol 4240 oil	137.0
Zinc oxide	3.0
Stearic acid	2.0
TMTD	1.80
MBT	.60
Sulfur	1.50
	395.90

Such a compound would be expected to have these properties:

Density (cured)	1.13
Tensile strength	1700 psi
Elongation	480%
Shore A hardness	60
Compression set	$\approx$ 20%
(ASTM method B: 22 hr at 70°C)	

Given the number of factors involved, it is doubtful that the first formulation would meet all requirements. If too costly, for example, increasing the black and oil loading would be the first adjustment. Good compounding comes with experience, that intimate knowledge of the properties of the materials being worked with and their interactions. Unless there is a particularly knotty problem, five or six different formulas should result in a specification-meeting recipe. Difficult compounding problems might involve a buyer's specifications written around or even provided by a present favored supplier's product, or certain products for which usage is very rough and the ambient conditions varied and uncertain, such as blowout preventers in oil drilling.

The somewhat elementary compounds exemplified above suggest how compounders may tackle their problems. Compounding problems in practice can be extremely varied. For example, problems the author has encountered have involved reducing the flow of gases through a stretched rubber membrane and blocking oil stain migration in products made of layered compounds.

Much help in solving problems can often be found in the literature. Outstanding here is *Rubber Chemistry and Technology*, published quarterly by the Rubber Division of the American Chemical Society, and especially the yearly *Rubber Reviews*. Some additional sources are listed below.

BIBLIOGRAPHY

1. *Science and Technology of Rubber*, F. R., Eirich, Ed. Academic Press, New York, 1978.
2. *The Vanderbilt Rubber Handbook*, 13th ed., F. F. Ohm, Ed. R. T. Vanderbilt Co., Inc. 30 Winfield St., Norwalk, CT 06855, 1990 (privately published).
3. *Basic Compounding and Processing of Rubber*, H. Long, Ed. Rubber Division, American Chemical Society, Washington, DC, 1985.
4. *Natural Rubber Formulary and Property Index*, Malaysian Rubber Producers Research Association, Brickendonbury, Hertford, England, 1990.
5. *Elastomers: Criteria for Engineering Design*, C. Hepburn and R. J. W. Reynolds, Eds. Applied Science Publishers, London, 1979.
6. *MRPRA: Third Rubber in Engineering Conference* Malaysian Rubber Producers Research Association, Brickendonbury, Hertford, England, 1973.

17

Compound Development—Part 2

As shown in Chapter 16, much compounding is done as an art based on knowledge of materials at hand, equipment available, product demands, and experience. In addition, repeated laboratory trials may be needed, to refine a compound until it is satisfactory. If opportunities for such testing are limited, general methods to speed up the process are welcome. Two possible aids come to mind: the use of computers and the use of statistically (as opposed to pragmatically) designed experiments.

Compounding by computers has a nice ring to it, and certainly calculators and computers can cut to a fraction the time required for the mathematical chores involved in determining costs, densities, test results, etc. A logical and great step forward would be to use computer programs for compounding itself. One difficulty is that a suitable data base containing the properties assigned to the various rubbers by different ingredients, does not exist. Standardization activities by the ASTM, other national standardization bodies, and the ISO may be moving us closer to the arrival of such a data base, which will permit the writing of the necessary software. The computer does allow compounding to be improved and shortened by making powerful statistical tools more readily accessible.

Many compounders feel the impingement of statistics in the areas of statistical quality control (SQC) and statistical process control (SPC). For those unfamiliar with these concepts, a brief example may be useful.

Consider a factory making all rubber molded footwear and having as one criterion of quality the stretchability of the upper. This might be measured by the 100% modulus, the force required to extend a sample (cut from the upper) to double its original length. Too low a modulus and the rubber has a tendency to slip off, too high and it is difficult to pull on over the shoe. Over a period of time when production appeared normal, with no processing or buyers' complaints, perhaps 100 samples might be collected and tested for this property.

The results, when plotted as a frequency distribution, probably will show a bell-shaped pattern, called a normal distribution. From these values we compute the mean (average), the range (highest minus lowest), and the sample standard deviation. The latter is a common measure of variability defined as:

$$S = \sqrt{\frac{\Sigma (x - \bar{x})^2}{n}}$$

where n is the sample size, $\bar{x}$ is the arithmetic mean, and Σ indicates summation of the quantity $(x - \bar{x})$ for all values of x. Unfortunately, if we are interested in the variance of all the observations, not just that of our sample, the quantity S does not provide an unbiased estimate. For this reason, ASTM publication STP 15D advocates use of a corrected estimate of the population standard deviation, replacing n in the denominator by $n - 1$. This has a negligible effect on S when the sample size is large, but it should be done.

If the values are normally distributed, approximately 68% of the modulus values will be within one standard deviation of the mean, 95% within two standard deviations, and 99.7% with n three. If the sample size is large enough, we can use $\bar{x}$ as the mean; we then expect at most one value in 300 (3 × (100 − 99.7)) to be outside the interval $\bar{x} \pm 3S$. If that is not the case, there may be reason to suspect that some condition is upsetting the process and changing the quality. Steps are then taken to right the situation. This kind of monitoring is statistical quality control (SQC).

In practice, of course, there are refinements. Average and range values for a set of routinely drawn samples are devised that relate to the overall average and standard deviation. Warning limits may be placed on the control chart used for monitoring the process so corrections can be made promptly.

The same kind of thinking may be used in setting up process controls. Instead of 100% modulus, the value monitored might be weight in grams of the uncured shoe. This control, tied to process conditions, is called statistical process control (SPC). Some buyers are now demanding SPC as well as SQC data. This may present competitive hazards if the confidentiality of the former is not strictly respected by the buyer. For example, a carbon black producer might learn that his competitor has changed to a more economical feedstock from SPC data, yet the SQC data will show no change.

To cope with these quality demands the compounder must establish quality limits early, and this is made easier if he can use materials that regularly are received with SQC data. This gives nore insurance that if a compound develops trouble, it will not be because of excessive variation in a compounding ingredient. If the customer has a specification, the compounder must be sure that the normal variation in properties of his compound do not exceed the limits set by the buyer. To be confident that his test values are accurate. the compounder can check them against the values for a standard compound (e.g., a carbon black test formula) and their repeatability and reproducibilty against the values given by the ASTM for the test in question. Compounders tending to innumeracy, like the writer, can find many statistical tools presented in simple fashion in M. J. Moroney's *Facts from Figures* [1].

Statistical methods for experimental design do not appear to be used by many compounders, especially in smaller companies. Possibly this lack is grounded in fears that the methods are too complex to master, much like earlier prejudices against computers. Undoubtedly there are some statistical methods that are difficult to learn. There are, however, simple ones that are effective and use only a small amount of mathematics. Statistical methods are helpful for decision making when uncertainty is prevalent.

Perhaps more than most industries, the rubber goods producers have to deal with questionable test results. Road testing of tires for the mileage to be expected is an example. Speed, inflation pressure, quality of road surface, load, and many other factors influence tire tread wear. Only by a statistically designed experiment can a proper amount of confidence be placed in any test results. The most compelling reason to use statistical methods in industrial experimentation is that they provide the most information for the lowest expenditure of money.

Statistical methods can help solve questions like the following, which occur in rubber compounding.

1. Are differences that show up in test results significant, or might they be caused by inherent variation in the test itself?
2. How does one design experiments to answer specific questions?
3. What conclusions are justified by the figures, and the analysis of data?
4. How does one optimize formulations?

For example, a rubber factory that has a limited mill room capacity orders a portion of a high volume stock from a custom mixer. What tensile difference between batches custom mixed and those mixed in-house would indicate a significant quality difference?

Or we might want to have an experimental program designed to indicate areas in which test sample variability could be reduced in the mixing, curing, or testing process. The method used for such an assignment is called analysis of variance (ANOVA). The writer applied ANOVA for one laboratory and was rather surprised to learn that each of the three operations contributed equally to the variability in results.

In general the statistician appears to be more conservative than the layman in assessing test results. Most technologists would feel that if something happened four out of five times, it was significant. The statistican reserves this term for an event that occurs 95% of the time, the so-called 95% confidence limit.

To show how simple statistically designed experiments can help, consider this production problem. A hose factory is having trouble with its 5/8-in., one-ply, braided water hose. Low burst strengths are showing up in more than 20% of its production. Low burst strength hose is considered to be scrap. Experience suggests three possible causes for this deficiency:

1. Poor quality braiding yarn. The factory has two suppliers, A and B.
2. Solids content of the rubber cement over the braid too high. Try regular and low solids cement?
3. The problem may be related to differences in the two work shifts. It is believed the first shift, 8 A.M to 4 P.M., may monitor yarn tension more closely than the second shift, 4 P.M. to 12 midnight. Irregular yarn tension can cause low pressure bursts.

To investigate, we could use what is called a four-run, two-level factorial design. This simply means that there are 2^2 experiments or runs, the two levels could be regular solids cement and low solids cement, shift 1 or shift 2, and so on. In our case the variables and levels are:

1. Yarn, A and B
2. Solids, regular and low
3. Shifts, 1 and 2

Four experiments are scheduled and we line up their conditions as follows:

Expt. 1. Shift 1, yarn B, regular solids
Expt. 2. Shift 2, yarn A, regular solids
Expt. 3. Shift 1, yarn A, low solids
Expt. 4. Shift 2, yarn B, low solids

On completion of the trials, we have the following results:

Expt. 1.   16% scrap
Expt. 2.   26% scrap
Expt. 3.   28% scrap
Expt. 4.   18% scrap

The analysis is simple. For each variable we calculate the percentage of scrap at the two levels. Thus we have:

Yarn A (26 + 28)/2 = 27% scrap
Yarn B (16 + 18)/2 = 17% scrap
Shift 1 (16 + 28)/2 = 22% scrap
Shift 2 (26 + 18)2 = 22% scrap
Regular solids (16 + 26)/2 = 21% scrap
Low solids (28 + 18)/2 = 23% scrap

In doing this analysis for each variable the other variables were ignored. This is because the variables are balanced. In averaging the effect of one variable, the other two variables cancel out because the two numbers in the average include both levels of the other two variables.

The results are illuminating. We can point to a 10% reduction in scrap with yarn B, no difference due to shift changes, and a 2% increase in scrap due to low solids. Under normal circumstances most production people would conclude that yarn A is of inferior quality and should not be used further. The more thorough-going would want to know whether the differences were statistically significant. This could be determined by a statistical comparison of the means using an ANOVA approach. Such a method is given by Ott [2]. From these results, we might learn that although high scrap had been attributed to yarn from supplier A, using yarn from supplier B would not reduce the scrap to an acceptable level. Thus further tests like this should be done. Perhaps a stronger yarn, or straining the tube stock, might be beneficial.

The analysis above assumes that there are no interactions among the three factors, yarn, solids, and shift. If we did not want to make this assumption, we could do a full factorial design with eight (2^3) combinations, defined by the three variables. For example, shift 1 would be tested with each combination of yarn and type of cement. It might be found that lower solids cement penetrates a less tightly spun yarn more thoroughly, therefore providing better bonding and burst strength. The analysis of variance test is applicable to this situation as well.

The illustration above is a screening method that is widely used to determine the major variables affecting a result. A similar statistically designed experiment that often is used in compounding, is the seven-variable, eight-trial, two-level factorial design. Table 17.1 shows how the experiments are set up: x_1 through x_7 are variables, and $+1$ and -1 are simply different levels of the variable involved. For example, if sulfur is one of the variables, -1 might represent 2.0 parts and $+1$ 3.0 parts. In a design like this, the variables need not be ingredients. One variable might be curing temperature: 287°F for level -1 and 307°F for level $+1$.

There is an extensive literature on such designs. For example there are two-level screening designs that have 12 runs and 11 variables and a 20-run design that allows 19 variables. Designs of these types are described at length by McLean and Anderson [3]. Historically compounders have experimented by changing one variable at a time or by using the educated-guess method, where several formulations are prepared with the hope that one will satisfy the requirements. How much more efficient to use, say, an eight-run, seven-variable design, which lets you know which variables are significant, and with much more confidence.

The simplest analysis of such factorial screening experiments is to compare the average of the $+1$ levels and the -1 levels of the variable concerned, since the other variables cancel out. In this case there would be four values at each level to average. More sophisticated analyses of the data require such techniques as regression analysis and probability plotting.

The greatest utility of the multiple-run, multiple-level method is that it shows the main effects that are due to a variable change from level -1 to $+1$. We do

Table 17.1 Statistically Designed Experiment for Seven
Variables and Eight Runs, at Two Levels

Run	Variables						
	x_1	x_2	x_3	x_4	x_5	x_6	x_7
1	−1	−1	−1	+1	+1	+1	−1
2	+1	−1	−1	−1	−1	+1	+1
3	−1	+1	−1	−1	+1	−1	+1
4	+1	+1	−1	+1	−1	−1	−1
5	−1	−1	+1	+1	−1	−1	+1
6	+1	−1	+1	−1	+1	−1	−1
7	−1	+1	+1	−1	−1	+1	−1
8	+1	+1	+1	+1	+1	+1	+1

not know whether the change is significant unless some of the tests are replicated, to establish the variability of test results. Another weakness is that the interaction effect(s) between two variables is not shown. For example, if both sulfur and accelerator are increased in our example, curing rate will accelerate.

Determining what the two-factor interactions are and using statistical methods to predict compound properties involve large steps in complexity and calculating resources that can be mentioned only briefly here. Two-factor interaction effects are the most frequent interactions n rubber compounding. To determine both main and two-factor interaction effects, a resolution V factorial design is needed. The simplest design is an eight-trial, three-variable design, illustrated in Table 17.2. Data from the design can be analyzed by linear contrasts, normal probability patterns, or multiple linear regression.

The next step up is the use of response surface methodology, and this requires a computer. The earlier experimental designs answered questions like: Do paraffinic or aromatic oils lower hardness most? Using response surface methodology, however, predictions can be made. For example, the compound will have an abrasion index of 110 ± 3 i 42 parts of black and 5 parts of oil are used. Rubber technologists by now are familiar with the contour plots and response surfaces generated by such techniques, as usually received from ingredient suppliers. The whole theme of statistical methods used in rubber technology, including response surface methodology, is reviewed by Derringer [4].

In troubleshooting and certain other aspects of the compounder's work, it is useful to devise a simple cause-and-effect diagram. Suppose the goal is excellent quality and uniformity of a batch used in a critical application—perhaps a military rocket. The cause-and-effect diagram might appear as in Figure 17.1. Here the factors affecting the quality of the batch are given roughly in sequence of their occurrence Experience tells us what occurs when the cause is changed (too much milling time = softer stock) and its relative importance. The wrong size of batch would give less trouble if the difference is small, than say, leaving

Table 17.2 Statistically Designed Experiment,
Resolution V, 2^3 Factorial

Run	Variables		
	x_1	x_2	x_3
1	-1	-1	-1
2	$+1$	-1	-1
3	-1	$+1$	-1
4	$+1$	$+1$	-1
5	-1	-1	$+1$
6	$+1$	-1	$+1$
7	-1	$+1$	$+1$
8	$+1$	$+1$	$+1$

the accelerator out (wrong assay). If trouble occurs then, with this as a guide and values of the miscompounded batch referred to earlier, identifying the trouble is less difficult and the solution easier.

It is interesting to review briefly how the use of statistics has grown. The original concept of SQC appears to have been developed by Shewhart and others at Westinghouse around 1933–1934. It was seized on by military procurement personnel in 1940–1945 to assure quality in such war essentials as ammunition and armor plate. Small arms ammunition at that time was commonly given repeated 100% inspection to prevent possible panic if the bullets started misfiring in combat. After World War II the use of the techniques languished except in Japan, where the procedures were taught by Deming. That and related teaching resulted in a significant improvement in the quality of Japanese products, an improvement responsible to some extent for the acceptance in the U.S. market of Japanese imports.

One of the first applications of statistically designed experimentation was conducted at an agricultural research station in Rothamsted, England, early in

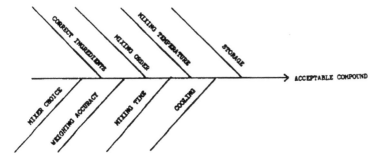

Figure 17.1 Cause and effect diagram.

this century. The decade 1955–1965 appears to have been particularly fruitful in the printing of the now classic works by Cochran and Cox [5], O. L. Davies [6], Mandel [7], and Box and Hunter [8]. Somewhat later, in 1968, the useful volume of Duckworth [9] was published.

Although the rubber goods industry can provide almost unparalleled opportunity for valuable utilization of these techniques, the number of contributors to this field seems small. One exception is G. C. Derringer who has published extensively on this subject. Two more of his publications [10, 11] provide a most helpful introduction to these methods for compounders.

Before leaving this subject, some general helps in problem solving might be mentioned. An early entry is B. S. Benson's approach to solving problems (he called it an alternative to brainstorming), the epistemological method of solving problems [12]. Certainly the six basic steps Benson outlines are applicable to a wide variety of problems. Another helpful technique is constructing the cause-and-effect diagram described by Ishikawa [13]. Although not directly related to compounding, familiarity with critical path techniques can be useful. Finally, another problem-solving technique is outlined in Warren Eberspacher's article *"Don't Jump to Conclusions"* [14].

REFERENCES

1. Moroney, M. J., *Factors from Figures*, Penguin Books, London, 1951.
2. Ott, E., *Process Quality Control*, McGraw-Hill, New York, 1975.
3. McLean, R. A., and Anderson, V. A., *Applied Factorial and Fractional Designs*, Dekker, New York, 1984.
4. Derringer, G. C., *Statistical Methods in Rubber Research and Development*, Rubber Chem. Technol., 61(3):377 1988).
5. Cochran, W. G. and Cox, G. M , *Experimental Designs*, 2nd ed., Wiley, London, 1957.
6. Davies, O. L., *Statistical Methods in Research and Production*, Oliver and Boyd, London, 1957.
7. Mandel, J., *The Statistical Analysis of Experimental Data*, Wiley-Interscience, New York, 1984.
8. Box, G. E. F., and Hunter, J. S., *The 2^{k-p} Fractional Factorial Designs, Part 1*, Technometric 3 (3): 311 (1961)
9. Duckworth, W. E., *Statistical Techniques in Technological Research: An Aid to Research Productivity*, Methuen London, 1968. Distributed in U.S.A. by Barnes and Noble, New York.
10. Derringer, G. C., *Statistical Process Control and the Design of Experiments*, Southern Rubber Group Meeting, June 15, 1984, Amelia Island, FL.
11. Derringer, G. C., *A Handbook for the Design and Analysis of Experiments*.
12. Benson, S. B., Let's Toss This Idea Up . . . , *Fortune Magazine*, October 1957.
13. Ishikawa, K., *Guide to Quality Control*, Unipub, New York, 1983, Chap. 3.
14. Eberspacher, W , Don't Jump to Conclusions, *INC.*, August 1980, p. 67.

18

Thermoplastic Elastomers

I. INTRODUCTION

With the rapid acceptance of thermoplastic elastomers (TPEs) in the past decade, they can hardly be ignored when reviewing rubber compounding; yet the field is so wide, any attempt to be comprehensive would require a separate volume. As a compromise this chapter is limited to a brief description of these materials where the conventional rubbers dealt with earlier are an integral part of their makeup. Rubber goods manufacturers would probably use only those containing rubber on introducing these polymers in their plants.

The rubbers described earlier were thermosets. By the application of heat and vulcanizing agents they lost their sticky, plastic, low strength qualities to become dry, firm, higher strength materials in an irreversible process. Thermoplastic elastomers have rubbery properties because they are a mixture of two domains or phases. One is a hard, high modulus phase such as polystyrene, the other is a soft flexible phase like polyisoprene. The hard domains replace the sulfur crosslinks in conventional vulcanized rubber although in TPEs the crosslinks are physical, not chemical. The hard domains give strength to these rubbers like carbon black does in regular rubber compounds.

Thermoplastic elastomers are not suitable for tires. The most abrasion-resistant one—polyurethane—has a low coefficient of friction so is rarely wanted when good rubber-surface traction is needed. For an industrial material, TPEs have enjoyed good sales growth They had an estimated 7% of the non-tire market in 1986 and the expectation was 12-15% by the mid-90's (1).

Several features have contributed to their success. They can be processed by high speed plastic techniques, they need no vulcanization and the scrap is reusable. They have desirable physical properties such as good elasticity and a wide range of tensile strengths, tear resistance and hardness, with proper compounding. The high growth rate was achieved despite a dichotomy- rubber goods manufacturers didn't realize plastic, not rubber, processing equipment was best for these materials and a misunderstanding by plastic processors of molded rubber goods features that were needed if they wished to enter that market. Despite such drawbacks these elastomers are now used in such varied items as automobile rub strips, fascia, footwear, sporting goods grips, and wire jacketing. Being thermoplastic they cannot be used where high temperatures are involved and do not have high solvent resistance in general.

In opting to use TPEs obviously both the advantages and disadvantages have to be closely considered. These considerations vary somewhat with the particular TPE chosen but can be summarized as follows.

Advantages:

1. Lower density than conventional rubbers.
2. Little or no compounding as most TPEs are processed as received.
3. Scrap can be recycled.
4. Simpler processing and shorter molding cycle times.
5. Some thermoplastic processing methods can be cheaper than those used for thermosets, such as blow molding.

Disadvantages:

1. TPEs do not process well on equipment designed for rubber. Plastic equipment such as longer barrelled extruders are necessary.
2. Material drying before processing is often required.
3. TPEs are rarely available below 50 Shore A hardness and there is a limited number below 80 Shore A.
4. They cannot be used where service temperatures exceed the melting point of the hard phase.

As thermoplastic elastomers were developed by various companies the latter devised acronyms for their products. Many of these terms were overlapping and are rarely used now. Since readers might find them in the literature a short list follows

TPR - thermoplastic rubber. Used by Uniroyal for an olefin blend rubber they inctroduced as the first thermoplastic rubber.

TPO - thermoplastic olefin, a TPE with two olefins such as polypropylene and EPDM.

DVA - dynamically vulcanized alloy -rubber plastic blend with some curative.
TPV - thermoplastic vulcanizate - similar to DVA.

A. Thermoplastic Types and Structures

Some TPE types are block copolymers, a term which has not been explained.
Copolymers of the monomers A and B for example can be of four arrangements.

1. Random, where the AB sequence is determined by chance such as AAB-
   BAAABBBBAB. SBR has a random choice sequence.
2. Graft, where the backbone is all A or B and the side chain is grafted on at
   certain points
   $$-AAAAAAA$$
   $$B_{BB}$$
3. Alternation, where there is a regular pattern of alternating repeat units.
   $$-ABABABAB-$$
4. Block. Here there is a long sequence of one monomer followed by a long
   sequence of another monomer.
   $$-AAAAAAAAAAAAAABBBBBBBBBBBBBBB-$$

A more detailed description of these arrangements is given by Hiemenz. (2).

Other types of thermoplastic elastomers which have large contents of conventional rubbers are (1) copolymer blends where olefins are blended mechanically and (2) alloys that have been characterized as having a synergistic reaction between polymer systems (3). The TPEs that are not reviewed here are the plyester, polyurethane and polyamide block copolymers.

II. STYRENE BLOCK COPOLYMERS

A. Introduction

Although the first TPE was thermoplastic polyurethane which appeared in the 1950s styrene block copolymers, which today enjoy a major part of the market, were not available commercially until the 1960s. Shell Chemical was the first to introduce these new materials using the trade name, Kraton, followed about three years later by Phillips Chemical who chose Solprene as their brand name.

These TPEs are made by a sequential anionic polymerization using an alkyl lithium inititator. The monomers used are styrene, isoprene, butadiene and poly (ethylene cobutylene). For structural illustrations these monomers can be identified as S, I, B, and EB respectively. The resulting polymer has two hard polystyrene ends with a rubbery mid-component. Accordingly the types available are identified as SBS, SIS, and SEBS. Most Kratons and Solprenes have styrene contents between 30 and 40%. For ease in processing both families have members which have been oil extended.

There are some basic differences between the Kratons and the Solprenes. The former has a linear structure, the latter a radial structure. The radial structure has a branched rubbery center block that is terminated at each end by a polystyrene block. This structure is achieved by using alkyl lithium initiators to polymerize the styrene, butadiene is then added which forms a partial midblock then linked with a multifunctional linking agent like silicon tetrachloride.

There are some basic differences between the linear and the radial styrene block copolymers. The radial polymers due to their more compact structure give a lower viscosity at equal molecular weights to the linear polymer, both in solution and melt viscosities. These lower values are valuable in the formation of adhesives. If commercial products of the two varieties are compared at the same styrene diene ratios the linear copolymer types have significantly higher tensile strengths and higher elongations at break.

Since their introduction there have been some changes in the producers of these materials. Phillips Chemical interests in Solprene were transferred to Housmex, A Novum Company. New producers in the U.S.A. include Firestonce Chemical and Enichem Elastomers Americas Inc.. Styrene block copolymers are also made in Europe and Asia, e.g. Vitacom from British Vita in the United Kingdom and Tufprene from Asahi in Japan.

B. Properties

Properties of these copolymers, especially the service temperature range, depend upon the T_g of the soft rubbery phase and the T_g and T_m of the hard phase where T_m is the melting point of the crystalline material. Values of these temperatures have been given by Holden (4). For the soft rubber phase T_g is $-90°C$ for polybutadiene and $-60°C$ for polyisoprene and poly (ethylene-cobutylene). For the hard phase T_g of polystyrene is 95°C and T_m 165°C (in compounds containing polypropylene). As the T_g of the soft rubber phase is reached the TPE is far too stiff and brittle for any service. Conversely as the T_g of the hard phase is reached and the hard phase approachs its melting point the material becomes too soft for use. As a first approximation a a sat sfactory service temperature range would be $-50°C$ to 95°C. Tensile strength can range from 500 to 4000 psi and hardness from 35 Shore A to 55 Shore D.

C. Compounding

Unlike polyester, polyurethane and polyamide block copolymers polystyrene block copolymers are not used as copolymers but are compounded with plastics, oils, fillers, resins and stabilizers. Thye can be sold as copolymers or as already compounded materials to the molder or other goods fabricator. A typical formulation using a Kraton D TPE for a sneaker or athletic shoe sole is given by Holden (5)

Kraton D Rubber	100
Polystyrene or styrene	
resin	80- 100
Filler	50- 80
Oil	90- 110
Stabilizer	0.8
Typical Shore A Hardness	45 - 55

Further details of such compounds are given in a Shell bulletin (6).

Adding polystyrene, polyprene or polyethylene to the block copolymer decreases the cost while increasing the hardness and resistance to ozone, especially if polyethylene or polypropylene is used. A reinforcing filler like carbon black does not increase tensile strength but does increase modulus, hardness, tear resistance and abrasion resistance. Other fillers such as clays and whitings decrease cost and increase hardness but lower tensile and abrasion resistance.

As might be expected adding oil lessens tensile, hardness and abrasion resistance although it increases flex life and melt flow. Again it decreases the cost of the compound. Naphthenic or paraffinic oils are preferred to highly aromatic oils which can plasticize the polystyrene domains.

A wide variety of resins have been used with these copolymers to modify hardness or adhesion. Resins used include coumarone indene resins, petroleum hydrocarbon resins and the more tackifying resins such as rosin esters and pinene resins.

Finally as these copolymers with an unsaturated backbone like isoprene are subject to oxygen and ozone attack Simpson points out that hindered phenols can be used with dithiodipropionate esters as useful antioxidants. (7). Resistance to ozone can be increased in the same way as with conventional rubbers by incorporating 25 to 30% EPDM.

D. Applications

The prime use of polystyrene TPEs is in soles for footwear, particularly athletic shoes, either as unit soles or direct molded soles. Advantages for this application are the high coefficient of friction that gives good traction and their abrasion resistance. Other uses are for sporting goods and electric wire and cables where their good low temperature properties are helpful. Although not in the rubber field a unique use is as a modifier for asphalt.

III. OLEFINIC BLENDS

A. Introduction

There is a large variety of products available that would fit this general description. For our purposes we will consider only those blends which are made up of a thermoplastic polymer (polyethylene or polypropylene) and an unvulcanized elastomer such as NR, SBR or EPDM. The blending is done mechanically. An attractive feature is that with care some of these blends can be made in house if an internal mixer is available. There are numerous manufacturers of these products. In the USA they include Monsanto Chemical, Teknor Apex, Ferro Corporation and A Schulman Co. European producers are British Vita, Esso, and C. W. Huls among others.

B. Properties

Because of the flexibility the manufacturer has in choosing the polymers and the proportion of each in the blend there is a vast array of such materials available. Accordingly anyone entering the feld should review the trade literature thoroughly to get the material with the properties needed.

Some generalizations can be made. Specific gravities are low, usually 0.88 to 1.04. Tensile strengths can vary from 400 to over 4000 psi. These blends are hard with Shore A hardnesses from about 55 to 96. With higher hardness grades strength increases but compression set decreases. Service temperatures up to 125°C are possible.

These materials have rather low abrasion resistance but good traction and resilience. Properties such as tensile, tear resistance and compression set deteriorate as the temperature increases. Because of their composition resistance to aliphatic solvents is poor but they do resist water, acids and bases and have good weathering resistance. They can be painted for outside trim or other uses.

Occasionally these products are made with a small amount of a crosslinker. This is usually a peroxide curing agent and partially vulcanizes the elastomeric portion, (frequently natural rubber) to give an improved more uniform blend.

C. Compounding

Many of the grades do not require compounding and are extruded or molded as received. Certain properties can be enhanced such as weather or flame resistance by adding antioxidants or flame retarders.

D. Applications

A major use of olefinic blends is in the automotive industry where they serve as rub strips, fender extensions and other items. They also are made into wire and cable jacketing, tubes, and molded goods.

IV. ELASTOMERIC ALLOYS

A. Introduction

There are two types of elastomeric alloys. In both cases proprietary treatments have caused interaction between their components that have resulted in an improved product over the simple mixture.

The duPont company markets a product which is called a melt processible rubber (MPR). Melt processible rubber has been described as a highly plasticized single phase combination of a chlorinated polyolefin, an ethylene vinyl acetate copolymer and an acrylic ester often with a significant content of carbon black (8).

Another elastomeric alloy is a two phase compound with a polyolefin such as polypropylene or polyethylene for the hard phase and vulcanized rubber for the soft phase. EPDM or NBR is commonly used as the soft phase. At the time of writing marketing of such an alloy but using natural rubber is planned for Malaysia. The major supplier of two phase elastomeric alloys is Monsanto Chemical using the trade names Santoprene, Geolast and Vyram.

B. Properties

In continuous service these alloys should not be used over 135°C. Like most of the thermoplastic elastomers reviewed here they have good resistance to polar fluids but the character of the soft phase determines whether they have good resistance to hydrocarbon oils. Obviously with an NBR soft phase resistance is good, if polyisoprene is used resistance would be low. Crosslinking of the non plastic component does of course increase fatigue resistance, oil resistance and lessens compression set.

Depending on composition and the amount of vulcanization in the two phase system hardness runs from about 55 Shore A to 50 Shore D. Ultimate tensile strength can reach 3500 to 4000 psi. MPRs have a distinctly higher specific gravity at 1.2 to 1.3 compared to 0.9 to 1.0 for the two phase systems. With carbon black having a specific gravity of about 1.80 the higher specific gravity of the MPRs is understandable.

C. Applications

Like the other thermoplastic elastomers reviewed these alloys are widely used in the automotive industries for under the hood parts, boots, and other items. Another market is in the mechanical rubber goods area where they are formed into tubing, special architectural glazing seals, gaskets, etc.

REFERENCES

1. O'Connor, G. E. and Rader, C. P., Chapter on Thermoplastic Rubber, Vanderbilt Rubber Handbook 1990, p. 264
2. Hiemenz, P. C., Polymer Chemistry, The Basic Concepts, Marcel Dekker, Inc. 1984, p. 12
3. O'Connor, G. E. and Rader, C. P., Chapter on Thermoplastic Rubber, Vanderbilt Rubber Handbook, 1990, p. 266
4. Holden, G., Rubber Technology 3rd Ed., M. Morton Ed. Van Nostrand Reinhold, 1987 p. 470

5. Holden, G., Rubber Technology 3rd Ed., M. Morton Ed. Van Nostrand Reinhold, 1987 p.476
6. Bulletin SC: 198-83 Shell Chemical Co. Houston 1984
7. Simpson, B. D., Chapter on Thermoplastic Rubber. Vanderbilt Rubber Handbook 1978 p. 236
8. O'Connor, G. E. and Rader, C. P , Chapter on Thermoplastic Rubber. Vanderbilt Rubber Handbook 1990, p. 267.

Testing and Specifications

I. TESTING

Good compounding goes for naught if tests of compound suitability are poorly chosen and/or improperly done. This is not a treatise on compound testing, but certain almost elementary precautions must be followed if test results are to be useful and credible.

To start with, there should be no question about the identity and quality of the ingredients used. For example, SMR 5 is a standard Malaysian natural rubber grade whose color and appearance is often similar to another grade, SMR CV. Yet if SMR 5 is used instead of SMR CV, the raw compound will be stiffer and the rate of cure somewhat faster. So strict separation of rubber types and grades is essential. Materials that have been exposed to contamination or damage (e.g., bags of whiting that have gotten wet) should be discarded. Periodic inventories offer an appropriate time to get rid of any chemicals in a laboratory stock that have exceeded the manufacturer's suggested shelf life (assuming the materials were date-marked when received) or become contaminated, as well as any that cannot be identified. Certainly a stock of reference materials of known quality should be on hand—for example, the industry reference black IRB 6.

Appropriate mixing has as its goal the blending of all ingredients with the least loss in weight, satisfactory dispersion of the ingredients, the least time and energy consumed, and no precuring of the stock. Mixing losses can be minimized in a variety of ways. Using an accelerator in rod form rather than as a powder minimizes loss by dusting. Another device is to add liquid softeners or plasticizers contained in dispersible low melting polyethylene bags. The order of mixing is important, yet it is difficult to give comprehensive guidelines because of the wide range of ingredients used and variation in mixers employed. Even a compound with only seven ingredients can be mixed in over 5000 different ways if all ingredients are added separately. Some help in determining mixing order and methods can be obtained from ASTM standard methods for evaluation of natural and synthetic rubber. Although the formulas are bare bones test formulas, the methods do give the order in which ingredients are usually added, as well as batch sizes and some idea of mixing times. After mixing, the batch should be weighed to make sure mixing losses are acceptably low. Excessive mixing losses should be suspected if the weight of a mixed compound is less than 99.5% of the compound weight if the compound contained fillers. With pure gum stocks, losses should be less than 0.3%. A cut surface should also be examined to see if ingredients appear to be properly blended. This can be done more thoroughly using a low power hand lens or, if a more rigorous examination is warranted, observing microtomed samples under a low power microscope.

Once mixed, the stocks should be protected from contamination and held for an appropriate time before curing (the time interval generally is specified in the test method). If contamination is believed possible, the raw slabs can be covered with Holland cloth or, even better, stored in a closed container. Preferably the batches should be held in a room where temperature and relative humidity are controlled. Otherwise high storage temperatures can start neoprene stocks precuring; also stocks containing hygroscopic fillers can absorb moisture and influence stress–strain results. The current ASTM recommendation is for a temperature of $23 \pm 2°C$ and a relative humidity of $50 \pm 5\%$.

Testing equipment must be well maintained. Again some simple precautions can make a great deal of difference n the accuracy of the results. Scales should be tested regularly. The indicated separation of mill rolls can be checked by passing small narrow lead strips through the mill bite near each end of the roll and measuring the resultant thickness by a dial thickness gage. Mill roll, press, oven, and bath temperatures should be checked regularly with thermocouples. Tensile dies should never rest on wooden supports but on fabric or sponge to preserve their fine cutting edge. Some laboratories test the tensile die being used each day to make sure that it is nick-free as evidenced by the point of rupture being random in specimens tested for that purpose. Tensile testers can be tested for accuracy by using a spring device Angle abrasion testers should have their abrasive wheels sharpened at intervals to maintain their abrasive ability. Since certain polymers, like polybutadiene, have a tendency to "grease" such wheels, they should be restored by abrading other compound wheels before running a new set of compounds.

The best test for a compound is of course a service test. A prime example is the testing of tread compounds by road tests. Using whole or multipart treads, the actual wear resistance of the compound is measured against a known performer using statistically designed procedures. Other features such as dry and wet skid resistance can be determined, as well.

Tread wear is developed quickly by continuous driving on courses of varying abrasiveness, usually in low traffic areas. The wear can be determined by measuring loss of tread depth or, in the case of whole treads, by weight loss also. Opportunities for service testing nontransportation items can sometimes be found: steam hose may be tested in a steam line in the plant itself, athletic shoes given to sports teams to evaluate, belts provided a customer on a service life cost basis, and so on.

Unfortunately such "real-world" testing cannot be done in many cases. There is insufficient time, the testing is too expensive, or there are other obstacles. Accordingly, most compounds are tested by the methods and to the specifications of the national testing authorities or establishments or the International Standards Organization (ISO). In the United States the ASTM provides almost all rubber testing methods and specifications, many of which, after passing through the American National Standards Institute (ANSI), become American standards and serve as a base for ISO standards. Although not enjoying complete protection, rubber goods manufacturers whose products comply to ASTM or ISO standards have some protection against product liability suits. The inference is that complying products represent the state of the art. Such protection would be especially useful for such consumer products as hot water bottles and garden hose.

A third area of testing—simulated service testing—is done in the laboratory and can be very useful. Many of these tests are trade secrets and are not matched by test methods in the public domain. Typical perhaps is a mechanical device used to test tennis balls. Here a racquet mounted on a revolving arm strikes in sequence a sample lot of balls over a period of time. Balls not maintaining a certain amount of resilience are collected as they fail to bounce over a barrier. Quality can be estimated from the average life of the units in the sample and the range.

Some 40 years ago a series of lectures by various experts on rubber technology was given at Newton Heath Technical School in England. In one lecture it was stated, "To design a laboratory test is easy, to determine the exact conditions of service is difficult, to combine the two and evolve the ideal service test is unknown." Even today this rings very true.

Few methods for specifying rubber characteristics would embrace most service conditions. One, again more than 40 years old, involves three basic properties and 12 coefficients [1]. The basic characteristics are tensile strength, resilience, and stiffness. The coefficients are the variation of these properties with temperature, speed, stress, and life (aging, physical fatigue, and chemical deterioration). The method is given in Appendix 5. It has been further described [2] as follows:

For any one application, measurement of the basic properties under the conditions of service will enable the amount of experimental work to be correspondingly reduced. Plastic characteristics can be represented by one property (viscosity) which will have coefficients related to the variables previously mentioned. It should be emphasized that the *fundamental* physical properties are stiffness and viscosity, while resilience is a composite of these properties and tensile strength falls far short of the intrinsic strength of the material and is simply a "practical estimate under the conditions used with a particular testing method."

It might be useful now to review briefly the use and importance of some common rubber tests in determining serviceability of compounds under stress and conditions of changes in stress and temperature.

Tensile tests are not very useful when predicting the behavior of a rubber compound in service. There is one noteworthy exception to this—bungee cords, where quality of the rubber rope is a matter of life or death. Incidentally the manufacture and nonsporting use of bungee cords are interestingly described in *Rubber Developments* [3]. Tensile tests do have some value as a quality control test in the manufacturer's plant as a rough measure of adequate dispersion.

If there is a first step (after selecting the polymer) in determining the suitability of a compound for a specific applications, it is probably hardness. This is often found on the basis of trial and error. Here is should be remembered that Shore durometers have their greatest utility as control testers; for precision work, hardness should be tested by a tester as specified in ASTM D1415 (identical to ISO 48) and the hardness given in international hardness degrees. Several specific properties needed from rubber in service are related to hardness. For example, work in the rubber laboratories of Imperial Chemical Industries showed that DuPont abrasion is primarily related to hardness in natural rubber compounds, provided a minimum tensile strength is maintained.

Tear resistance s required in many applications (e.g., footwear, belting) and some molded items such as chicken plucker fingers. Tear resistance is highest with crystallizable rubbers such as natural rubber, where strain induces crystallization and the crystals impede the tearing. For more confidence in the applicability of the results, testing should be performed with and perpendicular to the grain.

One of the most fundamental properties of a rubber compound is its resilience. Resilience is usually measured by noting the amount of rebound by a pendulum striking the rubber compound compared to the pendulum's initial position. There is a direct relationship between resilience and hysteresis:

$$\text{hysteresis} = \frac{(100 - \text{resilience}) \times \text{energy input}}{100}$$

In few cases in service does a half-cycle of stress, an impact, occur. If, however, the frequency of the impacts is increased until there is almost zero time between impacts, the condition of continuous vibration is approached. This is a frequent service requirement (e.g., for engine mounts), and tests selected should reproduce service conditions in order to predict performance satisfactorily. It is here that tests for determining such conditions, which include fatigue, heat buildup, abrasion, and flex cracking, multiply. They embrace some widely used testing devices like the Goodrich flexometer and the Akron angle abrader, and probably scores of private test methods.

If any service of this kind the heat developed follows the formula:

$$h = \frac{(100 - r)E}{100}$$

where r is the resilience and E is the energy input. The energy input varies with the service. It can be constant energy (the impact of a tire stud revolving at constant speed), constant load (the solid tire on a forklift truck running unloaded), or constant deformation (a belt).

Naunton and Waring [4] devised the following relations in these energy inputs:

$$\text{constant energy } H_1 = \frac{(100 - r)E}{100}$$

$$\text{constant load } H_2 = \frac{(100 - r)(F^2/2S)}{100}$$

$$\text{constant deformation } H_3 = \frac{(100 - r)(Sx^2/2)}{100}$$

where F is force, S is stiffness, and x is deformation.

The trouble is that in service the demands do not fit exactly into these three categories; instead, a given demand may call for tests in two or even three categories, depending on the application. Finding answers is made more difficult because of heat transfer and radiation.

In selecting tests for a compound, the magnitude of the stress, the frequency of a fluctuating stress, and the temperature at which the part operates must be of concern. In most cases these properties are difficult to come by, but the more specific this information, the easier the formulation of a satisfactory compound.

Rubber compounds under stress stiffen, and their resilience increases. Cassie et al. [5] were perhaps the first to find that when test specimens were flexed and not allowed to return to zero elongation, time to break of the sample was 4–10 times that when the samples were allowed to return to zero elongation. Some generalizations can be made about behavior after repeated stressing. If the stressing is to a constant minimum elongation, hysteresis declines, as do tensile strength and modulus. Unlike metals, rubber has a "memory"—resilience, for example, increases with repeated stressing in the pendulum rebound test, synthetic rubbers usually reaching their maximum before natural rubber.

Besides the amount of stress applied, we consider its frequency and the effect of temperature. The frequency can vary from very long period loading effects (e.g., the creep of a building mount to alternating or fluctuating cycles of stress—say, over 5 cycles/sec), or single loading cycles of long duration (e.g., determination of compression modulus). Again these types of stress fluctuation may mingle. For example, as a pneumatic tire increases in speed when it is driven against a drum, recovery of a tread stud is not completed in one revolution and the creep phenomenon takes place.

There are differences between static and dynamic properties. For example, it is claimed that the ratio of dynamic to static modulus, about 1.1:1 [6] for a soft natural rubber compound, is insensitive to frequency at usual frequencies (up to about 150 cycles/sec) and normal temperatures. "Dead" stocks (e.g., those with low resilience) usually have a high ratio of dynamic to static modulus. Both tensile strength and modulus show increases with rate of loading, such changes being most consistent with modulus.

Long-term loading effects are usually reported as permanent set, creep, or stress decay. In these tests it is most important that the experimental conditions be clearly defined so that the true change, the internal flow in the rubber, is measured with frictional slippage excluded as much as possible. It is the viscous component in a rubber compound that causes creep, and probably compression set at 70°C is most useful in predicting it.

In the design of many rubber parts, it is important to realize that creep or stress decay will occur over the entire life of the product. As a first approximation, one can assume that a logarithmic law holds for loss of both stress and creep. Permanent set tests, determined after the compound has been elongated at 100–200% for one week, should also give useful information on the creep that may be expected.

For any compound there are two limits to serviceability temperature-wise: (1) the low temperature at which the compound rapidly becomes harder, loses flexibility, and finally becomes brittle, and (2) the upper temperature limit, occurring when the compound becomes so soft that it is aging rapidly and cannot continue to function. The lower point can be determined by several tests (resilience at low temperature, torsional modulus at lower temperature, brittle point, etc.). The resilience of rubberlike materials generally passes through a minimum with reduction in temperature. This minimum, which occurs in the temperature resilience curve, varies from rubber to rubber. Some data are given in Table 19.1 [7].

As the temperature becomes lower, tensile strength, modulus, and permanent set increase, and elongation decreases. In compounding for low temperatures,

Table 19.1 Resilience Versus Temperature

Temperature of ball (°C)	Resilience (%)	
	Natural rubber	Neoprene
100	83	68
25		63
− 20	24	3
− 40	4	17
− 100	70	

Source: Data abstracted from Ref. 7.

however, hysteresis may cause the internal heat developed in a rubber to warm the part well above the ambient temperature. This may mean that a part can serve in a dynamic application where some conventional low temperature tests would indicate it to be unsuitable. In general, as far as polymer choice is concerned, good freeze resistance is incompatible with oil resistance.

Only general aspects can be considered with respect to high temperatures in service. As the temperature increases, tensile and tear strength decline, as do modulus and hardness. Elongation goes up and then declines, rebound resilience usually increases with temperature, and compression set increases sharply.

In compounding there are compatible properties, incompatible properties, and some properties that seem to have no relation to others. "Compatible" is used in the sense that if one property is improved, another property also improves. When incompatibility is present, one property improves, another declines. The third condition is "no apparent relationship between two properties." As a general rule, rubbers that have the most oil resistance are low in elongation, have poor resistance to freezing, and possess the worst electrical characteristics (i.e., electrical conductivity is relatively high, whereas most people think of rubber as an electrical insulator). Polymers with good elastic properties, such as high resilience, have relatively poor oil resistance and excellent freeze resistance.

A few general rules in dealing with compatible and incompatible properties are as follows.

High abrasion resistance is compatible with high hardness.
High hardness is not compatible with high resilience.
High tensile values have only haphazard relation to abrasion resistance.

In compound testing three gnawing questions are almost constantly present: How meaningful is this result? Have I really got an improved compound? Are the results I'm getting accurate? There are ways those questions can be answered. At periodic intervals a few batches, say four to eight, of the most commonly tested compounds can be tested. From the average, range, and standard deviation of the results from the various tests, one can develop a number that constitutes a meaningful difference for this type of compound. Although the results are specific to that one compound, they do give benchmark approximations for other compounds. Answering the second question can be done in several ways. One of the easiest is to participate in round-robin testing sponsored by the ASTM group concerned or other bodies. Another involves cooperative testing with customers. Finally, results can be checked with an established and recognized testing institution. Again, many of the problems entailed can be solved by the application of statistical techniques. Particularly useful is a paper by Derringer, "Statistical Methods in Rubber Research and Development" [8].

II. SPECIFICATIONS

Specifying a rubber part is accompanied by difficulties that are not experienced when metal parts are designed. For example, a rubber part made in one plant using the identical compound to that used in another plant may behave differently in service depending on the way it was cured—compression molded perhaps in one plant, injection molded in another. Such difficulties are best overcome by having a performance specification, preferably citing a formula that has given satisfaction but not restricting the supplier to that example.

Obviously the specification must be based on manufacturing experience so that it is known that the specification can be met. Another criterion of a good performance specification is that the manufacturer be able to carry out the performance tests. There are certain tests for which this is impossible, then such determinations must be made by an independent outside body. For example, several companies in Florida and Arizona offer rigorous outdoor tests for weathering that could not be performed on the premises of most companies. Finally it should be realized that any specification must be open for revision if experience dictates a change. Specifications should not stand in the way of progress.

Performance tests by their nature are extremely varied, and it would be an exercise in futility to try and cover them here, which leaves us with specifications where standard test properties alone are called for. Over the years many classification systems have been suggested for rubber compounds. One of the most sophisticated, flexible, and popular of these is the line callout system of Recommended Practice J200 of the Society of Automotive Engineers, also issued by the ASTM as designation D2000. The full ASTM title is Standard Classification System for Rubber Products in Auto notive Applications. The core of the system is the line callout, an alphanumeric designation that lists the properties required by letter and number. The system identifies by each letter or number the property involved and its value.

The utility of having a classification scheme for automotive rubber parts is evident when it is noted that apart from tires, the average car has over 600 rubber parts. Often this specification is used for nonautomotive parts. In the United States the scheme has been approved for use by agencies of the Department of Defense.

The system, which requires 30 pages of text in the ASTM book of standards cannot be repeated here, but an illustration of how it is used should suffice. A line callout is a specification and might look like this:

M4BC607A14EO33

The classification is based primarily on heat and oil resistance, always of concern in automotive design, followed by hardness and tensile strength. In our example, these properties are indicated by the sequence BC607.

The first letter, B, indicates the type of compound in its resistance to elevated temperatures. B indicates that a test temperature of 100°C is to be used in determining heat resistance. At this temperature the change in tensile strength is to be not more than ± 30%, elongation loss not more than 50%, and hardness change not to exceed ±15 points after 70 hr at this temperature. The test temperature for A (the lowest heat-resistant type) is 70°C; B as noted, is 100°C, and the test temperature then goes up by 25°C intervals to 275°C for type J.

The second letter, C, indicates the class of the compound in terms of oil resistance. Oil resistance, or lack thereof, is measured by volume swell. The volume swell is that which occurs after 70 hr immersion in ASTM oil no. 3 at a temperature determined from those listed for type but not over 150°C (the oil is unstable at these temperatures). Class A has no requirement for volume swell; class B has a 140% maximum swell requirement. The volume swell permitted decreases through letter classes until class E, which allows only 10%.

An appendix to D2000 is useful to compounders. It notes the type of polymer(s) most often used in meeting the type and class requirements, but it is not mandatory. In our illustration chloroprene polymers are most often used to meet the BC requirement. The first digit represents the hardness (e.g., 7 for 70 ± 5, 5 for 50 ± 5). The next two digits represent the minimum tensile strength wanted. If the line callout starts with an M, SI units are indicated, and 06 would mean 6 MPa. If there is no M in front, the tensile strength is that based on inch-pound units and the minimum value required in hundreds is used. For example, 17 represents a requirement for a 1700 psi stock.

Paired with the material letters BC in our example and the hardness and tensile strengths are minimum ultimate elongation values given in D2000 (or J200). In our case the minimum ultimate elongation is 300%; if the callout were BC807 the minimum elongation permitted would be 100%.

From the above, the shortest callout would be MBC607 in SI units. However, in many cases such a specification is too loose; deviations are wanted, or additional requirements exist. These are accommodated by using grade numbers before the letter classification in combination with suffix letters and numbers.

Grade numbers presently run from 1 to 9 and are inserted after M (if used) and before the two-letter material code to provide the basic properties. If the basic properties are sufficient, the grade number 1 is used (e.g., M1BC607), and no suffixes are added. Since 1986, suffix letters or combinations thereof have been used to identify 17 additional tests including one for any special requirement (which must be specified in detail)

In our example we are interested in two tests: heat resistance, specified by suffix A, and fluid resistance to oils and lubricants, specified by EO. The two digits after the A and the two digits after the EO identify the test method to be used and the temperature at which the test is to be run.

Tables in the classification method identify the digits to be used with various tests and also with the temperatures Our illustration calls for A14; the 1 stands for ASTM method D573 to be used; if test tube aging for the same time were wanted, 2 would be used, meaning ASTM method D865. The next digit, 4, stands for the temperature of the test, 100°C. The 3 after EO means that the ASTM method D471 is to be used with ASTM oil no. 3 for 70 hr. The final 3 represents the temperature, 70°C, at which the test is to be conducted.

The line callout then refers to a material that has these properties by Table 6 for BC materials in D2000.

1. SI units are used.
2. The material has a hardness of 60 ± 5 Shore A units, the minimum tensile strength is 7 MPa (equivalent to 1015 psi), and there is a minimum elongation of 300%.
3. The heat resistance has been t ghtened over the basic requirement. The maximum change in hardness now permitted is +15, in tensile strength −15%, and in ultimate elongation −40%.
4. In resistance to ASTM #3 oil, the maximum change in tensile strength is −45%, in ultimate elongation −30%, and the maximum volume increase permitted is +80%.

In the way it condenses and codifies a mass of information, D2000 (J200) is a remarkable document and rivals the systems used in stock market reporting and in giving thoroughbred racing results. It is hoped that these paragraphs make the system understandable, but only a meticulous review of the standard, with its tables, will allow a compounder to use this document freely.

In complying with buyers' specifications, when these can be obtained, the compounder should be aware of the degree of interlaboratory reproducibility of the test methods used. In many cases the reproducibility between other laboratories may be two to three times the repeatability obtainable in the compounder's own laboratory. Obviously an appropriate margin from the specification limits must be maintained. Such reproducibility data are hard to come by, but the ASTM is now requiring reproducibility statements as their methods undergo review and revision. Although these statements are sparse at present, as they are multiplied, difference in laboratory results will be clarified and better understood. For a comprehensive overview of physical testing, a contribution by Conant [9] is recommended.

REFERENCES

1. *Rubber in Engineering*, Appendix 2, *The Principal Physical Properties of Rubber and Their Coefficients of Change*, His Majesty's Stationery Office, London, 1946.
2. *Rubber in Engineering*, His Majesty's Stationery Office, London, 1946, p. 13.
3. White, L., *Rubber Dev.*, *40*(4), 96 (1987).
4. Naunton, W. J. S., and Waring, J. R. S., *Trans. Inst. Rubber Ind.*, *14*, 340 (1939); *Rubber Chem. Technol.*, *12*, 845 (1939).
5. Cassie, A. B. D., Jones M., and Naunton, W. J. S., *Trans. Inst. Rubber Ind.*, *12*, 49 (1937); *Rubber Chem. Technol.*, *10*, 29 (1937).
6. *Rubber in Engineering*, His Majesty's Stationery Office, 1946, p. 75.
7. *Rubber in Engineering*, His Majesty's Stationery Office, 1946, p. 91.
8. Derringer, G. C., *Rubber Chem. Technol.*, *Rubber Rev.*, *61*(3), 377 (1988).
9. Conant, F. S., Physical Testing of Vulcanizates, in *Rubber Technology*, 3rd ed., M. Morton, ed., Van Nostrand Reinhold, New York, 1987, p. 134.

20

Waste Reduction and Disposal

I. INTRODUCTION

Waste in this context means physical waste that occurs in rubber plant processing. Its handling and disposal are important because of escalating environmental regulations. These matters would not seem to be a responsibility of the compounder, but his or her technical knowledge and responsibility for chemical usage in the plant make participation in factory compliance desirable if not essential. Waste could be defined as material that is not suitable for its original purpose and must be discarded or recycled. This chapter deals with U.S. regulations, but Canada and other countries have similar requirements. Presently there is much public and political pressure to dispose of used auto tires effectively, efficiently, and safely. It is hard to expect the development of biodegradable tires that could end up in landfills. More immediately the solution may be to use scrap tires as part of the fuel in cement kilns. Pound for pound, scrap tires have about the same energy value as coal.

Besides complying with environmental regulations, reduction of waste is simply a good management practice. It is hard to judge the amount of in-process waste and defective finished product in U.S. rubber goods production. A conservative estimate might be 5% of finished production. If net profit is 10% of sales, the potential increase in profit obtainable by reducing waste is quite attractive. Waste includes not only finished product found to be substandard at the plant but product returned by customers as defective. Companies prosper only with continued customer goodwill. Returns erode that goodwill and are especially painful when the product is returned at the vendor's expense.

II. PHYSICAL WASTE REDUCTION

Waste can occur with raw materials, with product in process, or with finished product. Examples are frozen mold release compound, improperly braided hose, and finished product with an oil bloom. To reduce waste significantly, it is necessary to have good product planning, to purchase efficiently, to study all operations that cause waste, and to devise remedies for waste reduction. Finally one must continually check the process to see that the cures are working.

A. Product Choice

One way to reduce waste is to have a limited number of product lines. One polymer masterbatch producer, Company A, got into serious difficulties by producing special masterbatches for larger customers without assurances that sales would continue for a long period. So whereas most producers supply, say, a 50-part oil, 75-part carbon black masterbatch, Company A might accept an order for 45 parts of black and 80 parts of oil. As one could imagine, an overstocked warehouse resulted, containing odd amounts of odd masterbatches that were difficult to sell. The basic question is, How few compounds do you need to serve your customers well and what is the minimum number of ingredients necessary to have such a compound roster? Both raw materials and compounded batches can deteriorate in storage, whereupon they must be discarded as waste. Compounding materials such as magnesium oxide can pick up moisture on prolonged storage and lose effectiveness. Compounded batches can become useless as a result of bin scorch or premature curing.

Compounds should be as lean as possible in number of ingredients and still do the job. It is hard to believe that 18 ingredients are needed to make a satisfactory laboratory tubing, as given in one supplier's literature. Over 40 carbon black grades appeared in a recent ASTM classification scheme for that product. There is a good deal of overlap in properties among carbon black grades, and even a mechanical rubber goods factory should not require more than six or seven. Just as there should be a limited number of ingredients, the number of compounds should be limited. The roster of a mechanical rubber goods plant the writer was associated with had over 700 on its roster, several of which were used only once or twice a year; this certainly seems excessive.

B. Buying Practices

Packaging of consumer goods has become excessive: witness such durable items as screwdrivers sold in blister packages. Industrial products fare somewhat better, but a skillful buyer can reduce waste by looking critically at the packaging of incoming supplies. Such items as natural rubber crates and empty carbon black bags can be difficult to dispose of effectively. In the case of natural rubber, purchasing of polyethylene-wrapped, vacuum-packed loads on skids might be an advantage. If using is high enough, carbon black might be bought in bulk or purchased as a masterbatch with SBR or EPDM, if there is heavy usage of black with those polymers. Use of masterbatches, dispersions, and wetted powders of such items as sulfur, coloring agents, and resins tends to reduce the packaging materials requiring disposal. There is a definite trend to a lower amount of packaging materials, and it is a growing help in reducing waste.

C. Operation Analysis

Flowsheets of a plant's operations are essential in determining where waste occurs. Waste generating points can be targeted, the reasons analyzed, and corrective measures taken.

Consider a plant making high quality water hose (e.g., 1¼ in. i.d., 2-ply horizontally braided hose). A block flowsheet for the process is shown in Figure 20.1. In cross section, the hose consists of an inner extruded tube, covered with a braid reinforcing ply, a thin layer stock, a second ply of braid, and finally the cover. In use the tube conveys the liquid and protects the reinforcing plies from it. The braid plies give strength to the hose, allowing the fluid to be carried at high pressures without rupture. The cover protects the braid from cuts and abrasions, its smooth surface makes for easier handling. The thin layer stock is compounded to flow easily into the interstices of the brain plies, preventing the yarns from fraying by rubbing against each other; it also binds the tube, braids, and cover into an integral unit.

Six materials are used in this process. They include the cover stock, the braiding yarn (often rayon), the insulation stock, and the cover stock. Incidental materials are a pole lubricant and wrapping cloths. In making the hose, the tube stock is extruded, usually with a lubricant such as graphite on the inside. Cut to, say, 50½-foot lengths, the tubes are pulled over steel mandrels 1¼ in. in outside diameter and perhaps 51 feet long. The 50½-foot tube length allows for trimming the ends after vulcanization to give a standard 50-foot length of finished hose. The poled tubes, as they are then called, may be passed through two horizontal braiders in tandem. A roll of insulating compound tape of the correct width is mounted between the two braiders. The tape unrolls onto the first braid and is anchored by the second braid immediately afterwards. After the second brain has been applied, the hose is placed on a hose making machine, where the cover stock of calendered sheet is rolled on. Finally the hose is placed on a wrapping machine, where it is tightly wrapped with wet cloths and possibly wet cloth tapes. This is done to bind all the components close together and thus form a firm, dense wall in curing. After wrapping, the hose is vulcanized in a horizontal

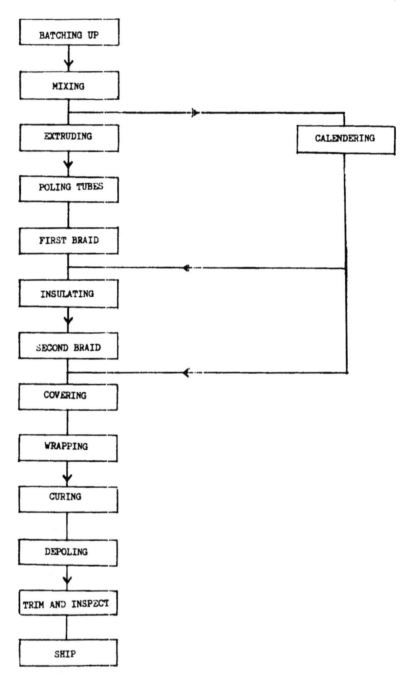

Figure 20.1 Braided hose process chart.

autoclave by open steam. After curing, the hose is removed from the vulcanizer, the wrapper(s) unwound, and the hose depoled. After cutting to the proper length (often couplings are added (at this stage), the hose is inspected and placed in storage or shipped.

A review of these operations shows several to be virtually waste-free. These include poling and depoling, braiding, and applying the insulation between the braid plies. The braiding operation does generate a minute amount of waste—the empty paper yarn tubes. Too trivial and fragile to be recycled as a paper product, their disposal is not difficult. In addition, the cloth wrapping and unwrapping processes are largely waste-free, since with proper care the cloths can be used again and again.

Definite, significant amounts of waste can occur in the storage, batching up, and mixing operations. More waste would be expected in the tubing and calendering operations for the insulation and cover stock. Finally, scrap is generated in trimming the finished hose and during any destructive testing (e.g., bursting test lengths to ensure that the product is strong enough for safe use with the pressures usually encountered).

The waste could be classified as nonrubber waste, such as excess packaging, materials and floor spills that occur in batching up; uncured waste, which might include poorly mixed stock or residues from the tubing and calendering operations; and finally cured waste, resulting from hose trimming and inspection. Handling this waste is addressed later.

In this hose-making operation the easiest way to determine where waste occurs and its amount entails the use of material balances. Batches going to the tubing and calendering departments would be weighed before mixing, and the stock actually delivered to those departments also would be weighed. The difference is the waste generated in the mixing operation. Similar calculations would be made by checking stock going to the tuber and determining the weight of extruded tubes going to the hose making department, in this instance the braiding section. In this fashion every operation would be studied, and any operations to be investigated further could be chosen from the Pareto principle, which realizes that a major result often is due to a minor cause. For example, it might be found that 75% of real estate sales were made by 25% of the sales force. Or, in a given plant, perhaps 70% of the rubber scrap is caused by 20% of the operations. Concentrating on the major causes in this way is logical. Cost of scrap reduction probably follows the cost of purification of a chemical, which rises exponentially as higher and higher assays are demanded.

A logical next step is to review environmental regulations and determine how compliance can be attained. Some general rules already apply. Any waste that can be recycled in a plant should be. Compositions of waste have to be known, and waste materials must be adequately separated. Any hazardous or toxic wastes must be plainly identified as such.

III. WASTE DISPOSAL AND THE LAW

High waste in a plant not only impairs the bottom line of net profits: if the waste is not handled according to prevailing environmental laws, major penalties may be incurred by the corporation and by the responsible individuals. U.S. factory managers have been jailed for disposing of toxic wastes illegally.

The environmental protection codes that govern a rubber plant's operations in the United States are found in five federal acts. These are:

Occupational Safety and Health Act (OSHA), 1970
Clean Water Act of 1977
Resource Conservation and Recovery Act (RCRA), 1976
Superfund Amendments and Reauthorization Act (SARA), 1986
Clean Air Act of 1992

Three of these acts do not involve the compounder directly. OSHA is concerned with safety in the workplace. Consider a hand building operation involving solvent adhesives. OSHA sets limits on the ambient concentration of dangerous solvents at the work location. For our purposes we would not ordinarily consider the solvent that escapes to the atmosphere to be waste. However if the operation involved very heavy solvent usage—for example, the manufacture of rubber surgical gloves from rubber cement, or electricians' gloves made by dipping—a solvent recovery system might be required.

Conforming to the Clean Water and Clean Air acts generally is not difficult for dry rubber operations, but latex goods production is a different matter. The compounder does, however, need a good understanding of the RCRA and SARA requirements. Two difficulties show up almost immediately. Unfortunately the administrators of RCRA and SARA cannot agree on listing materials as hazardous or toxic, and each group has its own lists. A further problem is that individual state EPAs can impose more severe requirements than the federal agency, and of course the rubber goods plant is bound by the requirements of the state in which it is located. It is generally recognized that the most restrictive state requirements are found in California.

A. The Resource Conservation and Recovery Act (RCRA)

RCRA regulations can and do change with time. The requirements given here have been in force since July 1990. The procedures a plant must follow in complying with the act are described in a booklet "Notification of Regulated Waste Activity" available from the federal EPA. It is up to the plant manager to determine whether a hazardous waste is handled in the plant, and if so, whether that particular waste is regulated by the RCRA. A company that handles a hazardous waste must advise U.S. and state EPAs on a standard form (8700-17) outlining its waste generation practices, whereupon a U.S. EPA identification number is issued. Figure 20.2 shows some of the information called for on Form 8700-17.

Like many government regulations, the RCRA regulations are lengthy, cumbersome, and ambiguous. Primarily a company must determine:

1. Whether the waste in question is solid waste, and if so whether it has been specifically excluded from the regulations.
2. Whether a given solid waste has been specifically listed as a hazardous waste.
3. Even a waste that is not listed may still be a hazardous waste due to its characteristics, and the plant manager is responsible for judging those characteristics against specified tests.
4. Nevertheless, it may be that the waste is exempt from the regulation, in which case there is no requirement to notify either the federal or the state EPA.

To begin at the beginning, a definition of waste is required. Federal EPA defines solid waste as any discarded material that is abandoned, recycled, or considered to be inherently wastelike. By this definition most rubber wastes would be called solid wastes. But again there is murkiness. Most compounders, for example, would consider latex goods plants to have liquid wastes, but some sludges from these sources there are considered to be solid wastes.

The list of hazardous solid wastes of the federal EPA appears in Section 261.31–261.33 of the Code of Federal Regulations (CFR). The list cannot be considered to be static; moreover, there are several hundred materials listed, most of which do not apply to rubber. But some common liquids used in rubber goods plants to appear, such as benzene and methyl ethyl ketone.

The next determination, referred to in item 3 above, is perhaps most prone to misunderstanding and expense. The plant manager is required to test the waste to determine whether it is ignitable, corrosive, reactive, or toxic unless it can be ascertained from its composition that it has none of these characteristics. Precise standards are given for these characteristics.

Some scrap, such as scrap from molding natural rubber swim fins, can be easily identified as nonhazardous. Because it is used in contact with the skin, the compound would already have met requirements for absence of reactivity, corrosiveness, and toxicity. The only other characteristic that might have a bearing would be ignitability. Under the EPA regulations this compound would not be classed as ignitable because (1) it is not a liquid, (2) it is not capable under standard temperatures and pressures of causing fire through friction, absorption of moisture, or spontaneous chemical changes, and (3) when ignited it does not burn so vigorously and persistently that it creates a hazard.

Please print or type with ELITE type (12 characters per inch) in the unshaded areas only

Form Approved. OMB No 2050 0028 Expires 10 31 91
GSA No 0248 EPA 07

ID – For Official Use Only

VIII. Type of Regulated Waste Activity *(Mark 'X' in the appropriate boxes. Refer to instructions.)*

A. Hazardous Waste Activity

1 Generator (See Instructions)
  a. Greater than 1000kg/mo (2,200 lbs.)
  b. 100 to 1000 kg/mo (220 - 2,200 lbs.)
  c. Less than 100 kg/mo (220 lbs.)

2. Transporter (Indicate Mode in boxes 1–5 below)
  a. For own waste only
  b. For commercial purposes
  Mode of Transportation
  1. Air
  2. Rail
  3. Highway
  4. Water
  5. Other - specify

3. Treater, Storer, Disposer (at installation)
  Note A permit is required for this activity; see instructions

4. Hazardous Waste Fuel
  a. Generator Marketing to Burner
  b. Other Marketers
  c. Burner - indicate device(s) -
  Type of Combustion Device
    1. Utility Boiler
    2. Industrial Boiler
    3. Industrial Furnace

5. Underground Injection Control

B. Used Oil Fuel Activities

1 Off-Specification Used Oil Fuel
  a. Generator Marketing to Burner
  b. Other Marketer
  c. Burner - indicate device(s)
  Type of Combustion Device
    1. Utility Boiler
    2. Industrial Boiler
    3. Industrial Furnace

2 Specification Used Oil Fuel Marketer
  (or On-site Burner) Who First Claims the Oil Meets the Specification

IX. Description of Regulated Wastes *(Use additional sheets if necessary)*

A. Characteristics of Nonlisted Hazardous Wastes. Mark 'X' in the boxes corresponding to the characteristics of nonlisted hazardous wastes your installation handles. *(See 40 CFR Parts 261.20 - 261.24)*

1. Ignitable   2. Corrosive   3 Reactive   4. Toxicity Characteristic
   (D001)   (D002)   (D003)   (D000)

(List specific EPA hazardous waste number(s) for the Toxicity Characteristic contaminant(s))

B. Listed Hazardous Wastes. (See 40 CFR 261.31 – 33. See instructions if you need to list more than 12 waste codes)

1	2	3	4	5	6
7	8	9	10	11	12

C. Other Wastes. (State or other wastes requiring an I D number   See instructions)

1	2	3	4	5	6

X Certification

I certify under penalty of law that I have personally examined and am familiar with the information submitted in this and all attached documents, and that based on my inquiry of those individuals immediately responsible for obtaining the information, I believe that the submitted information is true, accurate, and complete. I am aware that there are significant penalties for submitting false information, including the possibility of fines and imprisonment.

Signature	Name and Official Title (type or print)	Date Signed

XI. Comments

Note: Mail completed form to the appropriate EPA Regional or State Office. (See Section III of the booklet for addresses.)

EPA Form 8700-12 (07-90) Previous edition is obsolete.   –2–

Figure 20.2 Technical questions for EPA identification number.

B. The Superfund Amendments and Reauthorization Act

Achieving compliance under SARA may well require more effort and care than are needed for RCRA.

The first job of the plant operator is to determine whether the plant would be considered a covered facility within the meaning of the act. A covered facility has three characteristics:

1. It has more than 10 full-time employees.
2. It is in the Standard Industrial Classification Code 20-39.
3. It manufactures, imports, processes, or otherwise uses a toxic chemical.

The EPA makes a distinction between "process" and "otherwise use." The main difference is that processing is an incorporative activity. (Adding a small amount of ultramarine blue to a white compound to give a "whiter" white is an incorporative activity.) On the other hand, spraying a mold release fluid on a mold is an "otherwise use" activity. Plant management of a covered facility must report yearly to federal and state authorities the amount of listed toxic chemicals used in the last calender year and how these substances were disposed of if certain threshold values were exceeded. These values are 25,000 pounds for manufacture or processing and 10,000 pounds if the toxic material is otherwise used. Leaks must be reported immediately to state and local authorities.

Compliance with SARA requirements entails five steps.

1. Section 313, Title III, of the Superfund Amendments and Reauthorization Act (SARA) of 1986 has a list of chemicals considered to be toxic. The plant manager must determine whether the covered facility processes or uses any of the chemicals on that list. He or she must know not only all the chemicals used in the plant but also the composition of any proprietary products that are purchased. It is a little difficult to determine why some common products are on the SARA list; zinc oxide, for example, is widely used in adhesive compounds applies to the skin. Other commonly used materials on the list include titanium dioxide and glycol ethers. Since the list may change from time to time, probably by addition, the latest edition should be obtained from the local EPA office.

2. If the first step reveals that a listed toxic chemical is being used, it must be determined next whether the threshold limits for reporting have been exceeded. This can be done by simply by adding to the inventory of the chemical at the end of last year all purchases made during the current year, and subtracting this year's end inventory. Separate reports must be made if a compound is used both in processing and otherwise. Many plants must report on zinc oxide, which appears in so many compounds.

3. A more difficult problem is the third step—identifying the points of release of the chemicals for which reports are required. Releases are classified as three kinds: solid wastes, liquid wastes, and fugitive emissions to the air. It is

necessary to point out where each waste went. Solid waste must go to a licensed handler of that kind of waste. Liquid wastes similarly must be disposed of by a licensed waste handler such as a solvent/oil refiner and recycler. The "point of release for fugitive emissions to he air" indicates where in the process chain release occurred. Conformity to all these regulations demands a detailed flowsheet of the operation.

4. In addition to reporting where the release occurred, the amount must be estimated. There are four ways to estimate this.

Mass balance
Direct measurement
Engineering calculations
Emissions factors

Mass balance might be used to determine the release of zinc oxide. The amount of zinc oxide used during the year would be the amount held at the end of last year plus the amount received during this year less the stocks at the end of the current year. The zinc oxide content of product shipped could be calculated from the amount and type of compound used in product sold during the year. The difference in the two values, that used and the amount shipped out, would be the amount of waste.

Direct measurement might be used on, say, some methyl ethyl ketone that was discarded because it was replaced by another solvent.

Engineering calculations are more suitable for chemical process plants and are little used in rubber goods production.

Fugitive emissions of volatile materials to the ambient air are usually estimated using emission factors; the EPA has factors they will accept for these cases. Considerable help is given for the estimating process in an EPA booklet "Title III Section 313 Release Reporting Guide: Estimating Chemical Releases from Rubber Production and Compounding".

5. The chemical release inventory form for the preceding calendar year must be completed and returned by June 30. The form, EPA Form R, comes with instructions. Most of the form is concerned with identifying the waste generator and the waste handler who received the waste, as indicated by the portions of this form reproduced in Figure 20.3.

Presumably with further restrictions in mind, the EPA wants to know not only what toxic chemicals are being released but also whether reportees are eliminating toxic chemicals and/or reducing the use of the ones already used.

C. Coping with Compliance

Satisfying RCRA and SARA regulations and their revisions is not easy. Also there are other laws being promulgated which will be added to these and may involve compounders. Certain state and city ''right to know'' laws require that toxic and/or hazardous materials used in a plant be identified as such to employees and to local officials. Such information must be accompanied by instructions on how the materials are to be handled and what to do if a spill or leak occurs. In smaller plants the compounder may be involved in this teaching and information generating role.

Certain general rules can be followed to ease the compliance burden with respect to present and future regulations. These might be summarized as follows.

1. Keep highly accurate logs and records of all waste disposals. Backup for these records might be microfilm copies at an off-site location. Besides satisfying local authorities that compliance exists, there is another reason for keeping such records. More and more companies are trying to obtain certification through ISO standards 9000–9004. These standards are designed to ensure uniform quality and safety levels for firms engaged in manufacturing and selling goods and services on world markets. In quality and safety audits for such certification, waste disposal records probably will be reviewed.

2. Appoint an individual to be the contact person for environmental concerns and to act as liaison with the federal and state officials enforcing environmental regulations. The responsibilities for that position (include acquiring a thorough knowledge of all environmental regulations affecting the plant, issuing plant compliance rules and seeing that they are enforced, and to avoid unwelcome surprises, contacting environmental regulation authorities frequently. If new products are contemplated or if unusual problems arise, it is wise to retain a consultant.

3. All plant employees, no matter what positions they occupy, should be aware that there are severe penalties for violating EPA regulations. The message should be varied enough and repeated often enough to ensure that it is not ignored at any time.

4. EPA and certain other government agencies may ask for the disclosure of what a company considers to be trade secrets, which if made public would dull the firm's competitive edge. If a trade secret would in fact be disclosed by giving certain information, the EPA will allow so-called sanitized reports to protect such a company's advantage. It is necessary to have a strong case, though, to achieve this shielding—broad claims are not allowed.

5. Deal only with reputable, licensed waste handlers. For most small and medium-sized rubber goods factories, contracts to have all wastes handled by licensed haulers are generally advisable. Only large companies have the resources in talent, equipment, and finances to do the job in house.

Solid wastes are handled by waste handlers specializing in that type of waste. Liquid wastes are a different matter. Most rubber goods companies do not provide large quantities of liquid wastes, but there are noticeable exceptions,

(Important. Type or print; read instructions before completing form.)   Page 3 of 5

EPA FORM R **PART II. CHEMICAL SPECIFIC INFORMATION**	(This space for EPA use only)

---

1. CHEMICAL IDENTITY

1.1	☐ Trade Secret (Provide a generic name in 1.4 below. Attach substantiation form to this submission.)
1.2	CAS # ☐☐☐☐☐☐☐ - ☐☐ - ☐ (Use leading zeros if CAS number does not fill space provided.)
1.3	Chemical or Chemical Category Name
1.4	Generic Chemical Name (Complete only if 1.1 is checked.)

---

2. **MIXTURE COMPONENT IDENTITY** (Do not complete this section if you have completed Section 1.)
Generic Chemical Name Provided by Supplier (Limit the name to a maximum of 70 characters (e.g., numbers, letters, spaces, punctuation))

---

3. ACTIVITIES AND USES OF THE CHEMICAL AT THE FACILITY (Check all that apply.)

3.1	Manufacture:	a. ☐ Produce	b. ☐ Import	c. ☐ For on-site use/processing
		d. ☐ For sale/ distribution	e. ☐ As a byproduct	f. ☐ As an impurity
3.2	Process:	a. ☐ As a reactant	b. ☐ As a formulation component	c. ☐ As an article component
		d. ☐ Repackaging only		
3.3	Otherwise Used:	a. ☐ As a chemical processing aid	b. ☐ As a manufacturing aid	c. ☐ Ancillary or other use

---

4. MAXIMUM AMOUNT OF THE CHEMICAL ON SITE AT ANY TIME DURING THE CALENDAR YEAR

☐☐ (enter code)

---

5. RELEASES OF THE CHEMICAL TO THE ENVIRONMENT

You may report releases of less than 1,000 lbs. by checking ranges under A.1		A. Total Release (lbs/yr)				B. Basis of Estimate (enter code)	
		A.1 Reporting Ranges			A.2 Enter Estimate		
		0	1–499	500–999			
5.1 Fugitive or non-point air emissions	5.1a					5.1b ☐	
5.2 Stack or point air emissions	5.2a					5.2b ☐	
5.3 Discharges to water 5 3. ☐ (Enter letter code from Part I Section 3.10 for streams(s).)	5.3.1a					5 3 1b ☐	C % From Stormwater 5 3.1c
5 3.2 ☐	5.3.2a					5 3 2b ☐	5 3 2c
5.3.3 ☐	5 3.3a					5 3.3b ☐	5 3 3c
5.4 Underground Injection	5.4a					5.4b ☐	
5.5 Releases to land 5.5.1 ☐☐☐ (enter code)	5.5.1a					5.5.1b ☐	
5 5 2 ☐☐☐ (enter code)	5.5.2a					5.5.2b ☐	
5 5 3 ☐☐☐ (enter code)	5 5 3a					5 5.3b ☐	

☐ (Check if additional information is provided on Part IV-Supplemental Information.)

EPA Form 9350-1 (1-88)

Figure 20.3 Technical questions from EPA Form R.

EPA FORM **R**, Part III (Continued)   Page 4 of 5

6. TRANSFERS OF THE CHEMICAL IN WASTE TO OFF-SITE LOCATIONS

You may report transfers of less than 1,000 lbs. by checking ranges under A 1.	A Total Transfers (lbs/yr)			B Basis of Estimate (enter code)	C Type of Treatment/ Disposal (enter code)
	A 1 Reporting Ranges		A 2 Enter Estimate		
	0	1-499 500-999			
6.1 Discharge to POTW				6 1b ☐	
6 2 Other off-site location (Enter block number from Part II. Section 2.) ☐				6 2b ☐	6 2c ☐☐☐
6.3 Other off-site location (Enter block number from Part II. Section 2) ☐				6 3b ☐	6 3c ☐☐☐
6 4 Other off-site location (Enter block number from Part II. Section 2) ☐				6.4b ☐	6 4c ☐☐☐

☐ (Check if additional information is provided on Part IV-Supplemental Information)

7. WASTE TREATMENT METHODS AND EFFICIENCY

A. General Wastestream (enter code)	B. Treatment Method (enter code)	C Range of Influent Concentration (enter code)	D. Sequential Treatment? (check if applicable)	E. Treatment Efficiency Estimate	F Based on Operating Data? Yes No
7.1a ☐	7.1b ☐☐☐	7 1c ☐	7.1d ☐	7.1e ___%	7 1f ☐ ☐
7.2a ☐	7.2b ☐☐☐	7 2c ☐	7 2d ☐	7 2e ___%	7 2f ☐ ☐
7.3a ☐	7 3b ☐☐☐	7 3c ☐	7 3d ☐	7.3e ___%	7 3f ☐ ☐
7.4a ☐	7.4b ☐☐☐	7 4c ☐	7.4d ☐	7.4e ___%	7.4f ☐ ☐
7.5a ☐	7.5b ☐☐☐	7 5c ☐	7.5d ☐	7.5e ___%	7 5f ☐ ☐
7.6a ☐	7.6b ☐☐☐	7 6c ☐	7.6d ☐	7.6e ___%	7 6f ☐ ☐
7.7a ☐	7.7b ☐☐☐	7 7c ☐	7.7d ☐	7.7e ___%	7 7f ☐ ☐
7 8a ☐	7 8b ☐☐☐	7 8c ☐	7.8d ☐	7 8e ___%	7 8f ☐ ☐
7 9a ☐	7.9b ☐☐☐	7 9c ☐	7.9d ☐	7 9e ___%	7 9f ☐ ☐
7.10a ☐	7 10b ☐☐☐	7 10c ☐	7.10d ☐	7 10e ___%	7 10f ☐ ☐
7.11a ☐	7.11b ☐☐☐	7 11c ☐	7.11d ☐	7 11e ___%	7 11f ☐ ☐
7.12a ☐	7.12b ☐☐☐	7 12c ☐	7.12d ☐	7.12e ___%	7 12f ☐ ☐
7.13a ☐	7.13b ☐☐☐	7 13c ☐	7 13d ☐	7 13e ___%	7 13f ☐ ☐
7.14a ☐	7 14b ☐☐☐	7 14c ☐	7 14d ☐	7 14e ___%	7 14f ☐ ☐

☐ (Check if additional information is provided on Part IV-Supplemental Information)

8. OPTIONAL INFORMATION ON WASTE MINIMIZATION

(Indicate actions taken to reduce the amount of the chemical being released from the facility   See the instructions for coded items and an explanation of what information to include)

A. Type of modification (enter code)	B Quantity of the chemical in the wastestream prior to treatment/disposal			C Index	D Reason for action (enter code)
	Current reporting year (lbs/yr)	Prior year (lbs/yr)	Or percent change		
☐☐	_____	_____	_____ %	☐☐	☐☐

EPA Form 9350-1(1-88)

including plants that make rubber-tc-metal bonded parts where solvent degreasing is used to clean the metal. Such liquid wastes are often looked after by custom recyclers of solvents, oils and liquid chemicals. The recyclers usually require a waste stream survey before accepting the waste. Often document takes the form of a questionnaire in which the solvents or oils present are identified, as well as the amount of settled solids. the boiling ranges, and whether such toxic chemicals as arsenic or selenium are present. Usually liquid waste recyclers are reluctant to take material containing 10% or more solids. The recovered solvent can be returned to the plant or sold by the recycler, whose still waste can be sent to a cement kiln as fuel. Liquid waste recyclers can give good service to a plant, taking as little as a 55-gallon drum of material, recovering the solvent by distillation, and providing the client with the cradle-to-grave records required by EPA regulations for waste disposal If a minimum fuel value is present, some latex plant residues that are difficult to separate by distillation can be sent to a cement kiln for fuel.

Although there is a burgeoning amount of environmental protection books, magazines, and papers, it is difficult to find references solely concerned with rubber goods protection. Reference might be made to the Environmental Hazards Management Institute, a not-for-profit organization that offers training courses and other services dealing with hazard management and publishes a monthly magazine, *Hazmat World*. Those needing compliance guidance can find help in "Organizing for Compliance," by K. A. Roy (*Hazmat World*, February 1990).

Appendix One

Unit Volume and Density

Rubber compounds are rarely sold by weight. Long experience in designing rubber compounds and rubber compounds used with other materials such as metals or textile fibers has indicated that certain thicknesses or volumes are required for satisfying service life. Examples would be the thickness of a conveyor belt cover, the depth of tread on a tire, and the volume of rubber in a bridge bearing pad. Obviously the volume cost of a compound is important.

In the past specific gravity and density have been used interchangeably. However, since specific gravity cannot be unambiguously defined, density (mass per unit volume) should be used. Unless otherwise noted it can be assumed that the measurement is at 20°C.

In determining the volume cost of a compound the density is first calculated and then from the cost per pound or kilogram of the batch, the cost per unit volume is determined. Historically cost in the United States has been determined as cost per cubic inch. In countries using the metric system the unit is cubic centimeters. To agree with international standards (the SI system) the unit volume is kg/m^3 and has the same decimal position as grams per cubic centimeter.

The calculations are straightforward and can be followed from Table A1.1. The weight of each ingredient in grams is divided by its density in grams per cubic centimeter to get its contribution to the volume of the batch. The sum of the individual volumes is divided into the weight of the batch to get the compound density, in this case in grams per cubic centimeters. That density can be converted to pounds per cubic inch by multiplying by the factor 0.397. The cost per

Table A1.1 Density Calculation of Hose Tube Stock

Ingredient	Weight (g)	Density (g/cm³)	Volume (cm³)
Natural rubber (RSS #2)	100.0	0.92	108.70
Stearic acid	2.5	0.85	2.94
Zinc oxide	5.0	5.57	0.90
Antioxidant, hydroquinoline type	1.0	1.06	0.94
Hard clay	25.0	2.60	9.62
Sulfur	2.1	6.00[a]	0.35
TMTS	0.3	1.42	0.21
	135.90		123.66

[a] 6.00 used in place of 2.07 to allow for combination of rubber with sulfur in vulcanizate. Density (cured stock) = 135.90/123.66 = 1.10.

weight unit is obtained simply by multiplying the amounts of the various ingredients by their individual cost per weight unit, adding these values, and then dividing by the batch weight.

The preceding method using density values from suppliers' literature or reference books is satisfactory for most purposes. For very precise work it may not be accurate enough. The ingredient densities should be determined more precisely. Most references, for example, give the density of carbon black as 1.80, yet one supplier indicated that its range of furnace blacks varied 0.042 in specific gravity. If very high loadings of black were used, this might lead to significant differences. Other fillers or reinforcers used in large volumes may have the same problem. One method of determining the density of a pigment in rubber is by carefully mixing two rubber batches, the second containing exactly half the pigment used in the first batch but otherwise identical to the first. Weighing should be done very carefully and mixing done so that mixing losses are minimized. After mixing, both batches are weighed and samples from each vulcanized. The density of the vulcanizates are measured by a precise method such as using a pyknometer. The density of the material in question then is:

$$D = \frac{(W_1 - W_2)D_1 D_2}{W_1 D_2 - W_2 D_1}$$

where

W_1 = weight of first batch after mixing
W_2 = weight of second batch after mixing
D_1 = density of vulcanizate from batch 1
D = density of material being tested

Other materials that may have varying densities and should be checked in precise compounding include plasticizers and softeners, especially when used in large volumes. Small differences in density can be measured if the vulcanizates are tested in a density gradient tube.

Appendix Two

Standard Abbreviations

Many of the abbreviations given here have not been standardized by government regulations or decisions of standards-making bodies such as the International Standards Organization. Admittedly incomplete, this list is an attempt to bring together those abbreviations most often met by rubber compounders in the technical literature.

AR	Aromatic (referring to oil)
ASTM	American Society for Testing and Materials
B.P.	Boiling point
BAPFR	Bromomethyl-alkylated phenol formaldehyde resin
BDITD	Bis (diisopropylthiophosphoryl)disulfide
BIIB	Bromobutyl rubbers (ASTM designation)
BR	Polybutadiene rubbers (ASTM designation)
BMDC	Bismuth dimethyldithiocarbamate
CBS	N-cyclohexyl-2-benzothiazolesulfenamide
CdDEDC	Cadmium diethyldithiocarbamate
CDMDC	Copper dimethyldithiocarbamate
CIIR	Chlorobutyl rubbers (ASTM designation)
CR	Neoprene rubbers (ASTM designation)
CTP	N(cyclohexylthio)phthalimide

DBP	Dibutyl phthalate
DBTU	N,N^1-dibutylthiourea
DEBS	N,N'-dicylclohexyl-2-benzothiazolesulfenamide
DEG	Diethylene glycol
DETU	N,N^1-diethylthiourea
DIBS	N,N-diisopropyl-2-benzothiazolesulfenamide
DMBPPD	N-(1,3-dimethylbutyl)N^1-phenyl-p-phenylenediamine
DOP	Dioctyl phthalate
DOTG	Di-o-tolylguanidine
DPG	Diphenylguanidine
DPTH	Dipentamethylenethiuram hexasulfide
DTDM	Dithiodimorpholine
EPC	Easy processing channel
EDPM	Ethylene propylene rubbers, terpolymer (ASTM designation)
EPM	Ethylene propylene rubbers (ASTM designation)
ETPT	Bis(diethylthiophosphoryl)trisulfide
ETU	Ethylene thiourea
FA	Fatty acid
FEF	Fast extruding furnace
GMF	p-Quinone dioxime
GPF	General purpose furnace
GRS	Government rubber styrene (now obsolete term)
HAF	High abrasion furnace
HAF LS	High abrasion furnace, low structure
HI-AR	Highly aromatic (refers to oil)
HM	High modulus
HMT	Hexamethylene tetramine
HS	High structure
IIR	Butyl rubbers (ASTM designation)
IR	Synthetic polyisoprene rubbers (ASTM designation)
ISAF	Intermediate superabrasion furnace
ISAF HS	Intermediate superabrasion furnace, high structure
ISAF LS	Intermediate superabrasion furnace, low structure
LM	Low modulus
LS	Low structure

MBS	2-(Morpholinothio)benzothiazolesulfenamide
MBT	2-Mercaptobenzothiazole
MBTS	Benzothiazyl disulfide
MDB	Morpholinodithiobenzothiazole
MW	Molecular weight
MPFR	Methylolphenol formaldehyde resin
MT	Medium thermal
NAPH	Naphthenic (referring to oils)
NBR	Butadiene-nitrile rubbers (ASTM designation)
ND	Nondiscoloring
NDPA	*N*-nitrosodiphenylamine
NR	Natural rubber (ASTM designation)
NST	Nonstaining
ODPA	Octylated diphenylamine
PAR	Paraffinic (referring to oil)
PBNA	Phenyl-β-naphthylamine
phr	Parts per hundred of rubber
RA	Rosin acid
SA	Salt acid
SAF	Superabrasion Furnace
SAl	Salt alum
SBR	Styrene-butadiene rubber (ASTM designation)
SDEDC	Selenium diethyldithiocarbamate
SIR	Standard Indonesian rubber
SL-ST	Slightly staining
SMR	Standard Malaysian rubber
SPF	Superior processing furnace
SRF	Semireinforcing furnace
SRF HS	Semireinforcing furnace, high structure
SSR	Standard Singapore rubber
ST	Staining
TBBS	*N-t*-butyl-2-benzothiazolesulfenamide
TBTD	Tetrabutylthiuramdisulfide
TDEDC	Tellurium diethyldithiocarbamate
TEA	Triethanolamine

TETD	Tetraethylthiuramdisulfide
TMQ	Polymerized 2,2,4-trimethyl-1,2-dihydroquinoline
TMTD	Tetramethylthiuramdisulfide
TMTM	Tetramethylthiurammonosulfide
TMTU	Trimethylthiourea
TTR	Thai tested rubber
ZBDC	Zinc dibutyldithiocarbamate
ZBDP	Zinc-*o-o*-di-*n*-butylphosphorodithioate
ZDEDC	Zinc diethyldithiocarbamate
ZMBT	Zinc 2-mercaptobenzothiazole
ZMDC	Zinc dimethyldithiocarbamate

Appendix Three

Weights, Measures, and Conversion Factors

Although as early as 1866 an act of the U.S. Congress made metric weights and measures lawful in the United States, their usage has been limited in trade and commerce. With the growth of the International Standards Organizations (ISO) and the activities of certain standard bodies in this country, the use of a metric system, specifically the SI system (from System Internationale d'Unites) has increased. It is unlikely that the United States will embrace this system as thoroughly as the United Kingdom or Canada has done in the near future, but rubber technologists have to be familiar with SI units and their conversion from customary U.S. and British units and vice versa.

Presently the SI system has seven basic units. They are:

Quantity	Unit	SI symbol
Length	Meter	m
Mass	Kilogram	kg
Time	Second	s
Electric current	Ampere	A
Thermodynamic temperature	Degree Kelvin	K
Luminous intensity	Candela	cd
Amount of substance	Mole	mol

There are two supplementary dimensionless units which are rarely used in compounding or rubber testing.

Plane angle	Radian	rad
Solid angle	Steradian	sr

From these basic units a host of derived units have been adopted. Many of these, such as luminance, magnetic flux, and electric charge, are not often used by compounders. Some quantities that can occur in compounding or rubber testing are as follows:

Quantity	Unit	SI symbol	Formula
Angular velocity	Radian per second		rad/s
Area	Square meter		m^2
Density (mass)	Kilogram per cubic meter		kg/m^3
Energy (work)	Joule	J	N·m
Force	Newton	N	$kg·m/s^2$
Power	Watt	W	J/s
Pressure	Pascal	Pa	N/m^2
Quantity of heat	Joule	J	N·m
Stress	Pascal	Pa	N/m^2
Velocity	Meter per second		m/s
Viscosity, dynamic	Pascal-second		Pa·s
Viscosity, kinematic	Square meter per second		m^2/s
Volume	Cubic meter		m^3
Work	Joule	J	N·m

To keep relationships in the basic units despite the existence of very large or very small quantities, prefixes are used as multiplication factors. The approved SI prefixes follow. However, in calculations the prefixes are changed to scientific notation for ease in calculating.

Multiplication factor	Prefix	Symbol
1 000 000 000 000 000 000 = 10^{18}	exa	E
1 000 000 000 000 000 = 10^{15}	peta	P
1 000 000 000 000 = 10^{12}	tera	T
1 000 000 000 = 10^{9}	giga	G
1 000 000 = 10^{6}	mega	M
1 000 = 10^{3}	kilo	k
100 = 10^{2}	hecto[a]	h
10 = 10^{1}	deka	da
0.1 = 10^{-1}	deci[a]	d
0.01 = 10^{-2}	centi[a]	c
0.001 = 10^{-3}	milli	m
0.000 001 = 10^{-6}	micro	
0.000 000 000 = 10^{-9}	nano	n
0.000 000 000 000 = 10^{-12}	pico	p
0.000 000 000 000 000 = 10^{-15}	femto	f
0.000 000 000 000 000 000 = 10^{-18}	atto	a

[a]It is recommended these be avoided where practical.

Conversion Units:
Customary U.S. and British Units to SI

Multiply	by	To obtain
Inches	2.54	cm
Feet	0.3048	m
1/16 inches	1.5875	mm
Square inches	6.4516	cm^2
Square feet	0.0929	m^2
Cubic inches	16.387	cm^3
Cubic feet	0.0283	m^3
Ounces	28.3495	g
Pound (mass)	0.4536	kg
Pound force per square inch (psi)	6.894×10^3	pascal (Pa)

Note: This last conversion is the one used in converting such stress values as tensile strength and modulus values to the SI system. Because tensile strengths are usually over 1000 psi, it is customary to report these values in kPA.

Note 2. For ease in calculations many conversions to SI are given in scientific notation. For example, the conversion factor used to convert square inches to square meters is 6.451 600 E−04. E refers to the exponent and is followed by a minus or plus sign and then two digits. The two digits are the power of 10 by which the number must be multiplied to obtain the correct value.

TEMPERATURE CONVERSION

For most measurements the Practical Kelvin Temperature Scale of 1960 is used. However, the units in both Kelvin and Celsius temperature intervals are the same. Accordingly, a Kelvin temperature can be obtained by adding 273.15 to the Celsius temperature. Celsius temperature is obtained from Fahrenheit values by using the formula $C = 5/9 (F - 32)$.

CHANGING UNITS

Many times it is necessary to change from one set of units to another for consistency. Mistakes can occur but can also be avoided by using the following procedure. The quantity to be converted (e.g., yards/hr) should be written in the original numbers with all units noted in words or symbols (e.g., 30 yards/hr). Then treat the units like algebraic quantities that can be divided, multiplied, or canceled. The overall rule for changing the units is to multiply the original values by fractions that equal 1. Suppose we have to change 30 miles/hr to ft/sec. We know that 5280 ft = 1 mile, 1 hr = 60 min, 1 min = 60 sec. Then

$$\frac{30 \text{ miles}}{\text{hr}} \times \frac{5280 \text{ ft}}{\text{mile}} \times \frac{\text{hr}}{60 \text{ min}} \times \frac{\text{min}}{60 \text{ sec}} = 44 \text{ ft/sec}$$

The trick is that all of the multiplying factors have the value of 1 and are used so that the right units will be in the numerator and denominator. As a further example, suppose we want to convert the speed of a slow (?) crayfish—2 yards/day—to cm/sec. We know that 1 yard = 36 in., 1 in. = 2.54 cm, 1 day = 24 hr, and 1 hr = 3600 sec.

The calculation then is:

$$\frac{2 \text{ yards}}{\text{day}} \times \frac{\text{day}}{24 \text{ hr}} \times \frac{\text{hr}}{3600 \text{ sec}} \times \frac{36 \text{ in.}}{\text{yard}} \times \frac{2.54 \text{ cm}}{\text{in.}} = 0.0021 \text{ cm/sec}$$

Obviously conversion to another set of units would be more meaningful.

RECOMMENDED READING

Swartz, Clifford E. *Used Math for the First Two Years of College Science.* Englewood Cliffs, N.J.: Prentice-Hall, Inc., 1973, chaps. 1 and 2.
ASTM Metric Practice Guide, American Society for Testing and Materials, 1916 Race Street, Philadelphia, Pa. 19103

Appendix Four

Density and Unit Volume of Compounding Materials

Even among inorganic materials there are surprisingly wide ranges for density values in the published literature. This probably reflects differences in composition, if the material comes from a natural source; differences in methods of manufacture if it is man-made. As might be expected, ranges are greater if natural fats and oils or color pigments with different hues are involved. For brevity, the small amounts that are used and their ready availability from suppliers, densities for accelerators, and age resisters are omitted. For costing and design purposes the following figures should be satisfactory. However, if large volumes of an ingredient with a range of densities are involved (e.g., factice), confirmation of the density should be made in the laboratory or received from the manufacturer.

Material	Density (g/ml)	Unit volume (ml/g)
Aluminum powder	2.50	0.40
Aluminum silicate	2.60	0.38
Ammonium bicarbonate	1.59	0.63
Antimony sulfide (pentasulfide, antimony red)	4.12	0.24
Antimony trioxide	5.67	0.18
Asphalt	1.0	1.0

Material	Density (g/ml)	Unit volume (ml/g)
Barytes	4.50	0.22
Bentonite	2.50	0.40
Bitumen	1.04	0.96
Blanc fixe	4.30	0.23
Butyl rubber	0.91	1.10
Cadmium yellow (sulfide)	4.82	0.21
Cadmium oxide	8.15	0.12
Cadmium red (cadmium selenide)	4.30	0.23
Calcium carbonate	2.70	0.37
Calcium silicate	2.10	0.48
Carbon black	1.80	0.56
Carnauba wax	0.99	1.01
Castor oil	0.945–0.970	1.06–1.03
Ceresin wax	0.92	1.09
Clay	2.60	0.38
Cork (ground)	0.22–0.26	4.55–3.85
Cotton flock	1.54	0.65
Cottonseed oil	0.92–0.93	1.09–1.08
Diatomaceous earth	1.9–2.35	0.53–0.43
Dibutyl amine	0.77	1.30
Dibutyl phthalate	1.04	0.96
Diethylene glycol	1.12	0.89
Dibutyl sebacate	0.936	1.07
Dolomite	2.85	0.35
Factice	1.04–1.15	0.96–0.87
French chalk (talc)	2.80	0.36
Gelatin	1.27	0.79
Glue (bone)	1.27	0.79
Graphite, flake	2.30–2.72	0.43–0.37
Ground glass	2.4–2.6	0.42–0.38
Gypsum	2.31–2.33	0.43
Iron oxide (red)	4.5–5.1	2.22–1.96
Kaolin	2.60	0.38
Kerosene	0.81	1.23
Kieselguhr	2.20	0.45
Koresin	1.03–1.04	0.97–0.96
Lanolin	1.08	0.93
Lauric acid	0.833	1.20
Lead dioxide	9.375	0.11

Material	Density (g/ml)	Unit volume (ml/g)
Lead sulfate	6.92	0.14
Lime, hydrated	2.20	0.45
Litharge	9.30	0.11
Lithopone	4.30	0.23
Magnesia LC	3.65	0.27
Magnesium carbonate	2.22	0.45
Mica, powdered	2.60–3.20	0.38–0.31
Mineral rubber	1.01–1.05	0.99–0.95
Mineral oil	0.84–0.94	1.19–1.06
Natural rubber	0.93	1.08
Neoprene (G,W)	1.23	0.81
Nitrile rubber (med. acrylonitrile)	0.96	1.04
Oleic acid	0.90	1.11
Ozokerite wax	0.85–0.95	1.18–1.05
Palm oil	0.75–0.88	1.33–1.14
Paraffin wax	0.87–0.91	1.15–1.10
Petrolatum	0.815–0.880	1.23–1.14
Pine oil	0.925–0.942	1.08–1.06
Pine pitch	1.105	0.90
Polyisoprene	0.92	1.09
Rosin	1.07–1.08	0.93
Salicyclic acid	1.26	0.79
Styrene-butadiene rubber	0.94	1.06
Shellac	1.10–1.15	0.91–0.87
Silica	1.95	0.51
Soapstone	2.7–2.8	0.37–0.36
Sodium acetate $NaC_2H_3O_2$	1.528	0.65
Sodium acetate $NaC_2H_3O_2 \cdot 3H_2O$	1.45	0.69
Sodium bicarbonate	2.159	0.46
Stearic acid	0.85	1.18
Sulfur	2.05	0.49
Talc	2.7–2.8	0.37–0.36
Titanium dioxide (anastase)	3.88–3.90	0.26
Titanium dioxide (rutile)	4.10–4.20	0.24
Tricresyl phosphate	1.18	0.85
Triethanolamine	1.12	0.89

Material	Density (g/ml)	Unit volume (ml/g)
Ultramarine blue	2.30–2.50	0.43–0.40
Urea	1.33	0.75
Vaseline	0.815–0.880	1.23–1.14
Vermilion (HgS)	8.10	0.12
Whiting, Gilder's	2.65	0.38
Whiting (ground limestone)	2.65–2.70	0.38–0.37
Wood flour	1.25	0.80
Zinc carbonate	3.85	0.26
Zinc oxide	5.55	0.18
Zinc stearate	1.05–1.10	0.95–0.91
Zinc sulfide	3.9–4.2	0.26–0.24

Appendix Five

Principal Physical Properties of Rubber and Their Coefficients of Change

The mechanical properties of rubber under service conditions are functions of three or more independent variables and can thus be represented in terms of partial differentials of single variables, the other variables being kept constant, i.e., if $z = f(xy)$ where x and y are independent variables and functions of some other variable t, then

$$\frac{dz}{dt} = \left(\frac{\delta f}{\delta x} \right)_y \cdot \frac{dx}{dt} + \left(\frac{\delta f}{\delta y} \right)_x \cdot \frac{dy}{dt}$$

In order to obtain the total change in a property (z) under a given set of service conditions, an attempt is, in fact, being made to measure the total differential (dz/dt). This is made easier if, by suitable measurements, the contributions of individual factors are determined separately, other factors being kept constant.

Thus, to calculate the variation of tensile strength (W) in service,

$$dW = \frac{\delta W}{\delta T} \cdot dT + \frac{\delta W}{\delta n} \cdot dn + \frac{\delta W}{\delta A} \cdot dA \ldots$$

| Change in tensile strength with service | Temperature contribution | (Frequency) contribution | Aging contribution |

The principal properties, conditions, and coefficients are listed here for reference.

(a) Properties   *Symbol*
    (1) Tensile strength   W
    (2) Stiffness   S
    (3) Resilience   R
    (4) Viscosity μ

(b) Conditions
    (1) Temperature   T
    (2) Speed (frequency)   n
    (3) Stress   P
    (4) Life   A

(c) Coefficients

    (1) Variation with temperature (other conditions constant)

$$\frac{\delta W}{\delta T} \qquad \frac{\delta S}{\delta T} \qquad \frac{\delta R}{\delta T} \qquad \frac{\delta \mu}{\delta T}$$

    (2) Variation with speed

$$\frac{\delta W}{\delta n} \qquad \frac{\delta S}{\delta n} \qquad \frac{\delta R}{\delta n} \qquad \frac{\delta \mu}{\delta n}$$

    (3) Variation with stress

$$\frac{\delta W}{\delta P} \qquad \frac{\delta S}{\delta P} \qquad \frac{\delta R}{\delta P} \qquad \frac{\delta \mu}{\delta P}$$

    (4) Variation with life

$$\frac{\delta W}{\delta A} \qquad \frac{\delta S}{\delta A} \qquad \frac{\delta R}{\delta A} \qquad \frac{\delta \mu}{\delta A}$$

In many cases there exist physical terms for these partial differential coefficients, for example,

Thermoelastic effect: $\dfrac{\delta S}{\delta T}$

Heat aging: $\dfrac{\delta W}{\delta A}$

Thixotropy: $\dfrac{\delta \mu}{\delta h}$

Anomalous viscosity: $\dfrac{\delta \mu}{\delta P}$

Thermal softening, plasticizing: $\dfrac{\delta \mu}{\delta T}$

Index

Milton Keynes UK
Ingram Content Group UK Ltd.
UKHW020314111024
449327UK00040B/1086